D1442788

Seed to civilization
THE STORY OF MAN'S FOOD

A series of books in biology

EDITORS
Donald Kennedy
Roderic B. Park

COURTESY OF THE ROCKEFELLER FOUNDATION

Seed to civilization

THE STORY OF MAN'S FOOD

Charles B. Heiser, Jr.
INDIANA UNIVERSITY

W. H. FREEMAN AND COMPANY
San Francisco

Library of Congress Cataloging in Publication Data
Heiser, Charles Bixler, 1920-
  Seed to civilization.

  Includes bibliographical references.
  1. Agriculture—History. 2. Food. 3. Domestica-
tion. 4. Botany, Economic. 5. Nutrition.
I. Title.
S421.H4   630'.9   73-2949
ISBN 0-7167-0593-1
ISBN 0-7167-0594-X (pbk)

Copyright © 1973 by W. H. Freeman and Company

No part of this book may be reproduced by any mechanical,
photographic, or electronic process, or in the form of
a phonographic recording, nor may it be stored in a retrieval
system, transmitted, or otherwise copied for public or
private use without written permission of the publisher.

Printed in the United States of America

International Standard Book Number: 0-7167-0593-1 (cloth);
                                    0-7167-0594-X (paper)

10 9 8 7 6 5 4 3 2 1

to Dorothy

WHO HAS KEPT ME WELL FED

Contents

Preface

There are many problems facing mankind today—war, poverty, hunger, pollution, race relations. These problems are so interrelated that it is difficult to single out one as being more important than the others. Some people, however, think that hunger is the greatest, a problem that—until recently—many people didn't even know existed in the United States. Hunger certainly ranks as one of man's oldest problems. Except for the last chapter, however, this is not a book about hunger. Rather it concerns mostly the plants and animals that stand between man and starvation. The subject can be called ethnobiology, the study of plants and animals in relation to man. It in a sense concerns ecology also. Ecology is the study of organisms in relation to their environment. In this case our organism is man, and that part of the environment of primary concern to us is the plants and animals that provide his food. But it would be stretching a point to say that the primary emphasis of the book is ecological, and for that reason I have refrained from jumping on the current bandwagon and putting "ecology" in the title.

Man, frequently of necessity, has always had a great interest in food. Eating is the second favorite activity of many people, and for some it is the first. Those of us who are not hungry can make jokes about eating, but it is hardly a laughing matter to half of the human population. We don't know exactly how many hungry people there are in the world. Even if there were a general, accepted definition of hunger, it would still be difficult to arrive at satisfactory estimates for some parts of the world. I recall many years ago seeing a slogan in a Boy Scout manual, "eat to live, don't live to eat."* That too many people live to eat is perhaps reflected by the present emphasis on the development of low calorie foods in the United States.

In this book I begin with some consideration of the origin of agriculture and why man domesticated plants and animals. The bulk of the book is concerned with man's most basic food plants and animals, and covers where and when they were first domesticated as well as why and how they are used. I have, however, not hesitated to stray from the principal subjects from time to time when I have felt that the digression would be of general interest to my readers. There is, for example, some mention of the uses of plants and animals for purposes other than food. The plants treated include all of man's basic food crops. Sugar cane and sugar beet receive somewhat less attention than the others, for although sugar is an excellent source of calories, it is important for adding interest to our diet rather than in supplying substances necessary for health. Only one chapter is given to the discussion of animals, and attention is focused on those most important as food. As I am a botanist, some may think that I have neglected animals in favor of plants, but in defense I can point out that man gets all of his carbohydrates and nearly three-fourths of his protein from plant sources. Moreover, nearly all of the food we get from animals is in turn derived from plants. After all, life depends on photosynthesis; chlorophyll has been referred to as the green blood of the earth. The last chapter concerns current and future food problems and, perhaps, some controversial topics.

Teaching courses in economic botany and plants, animals, and civilization has given me some background for this book. Although I have conducted research into the origin and relationships of some food plants, I have never worked with the really basic ones that are my chief subjects here. Thus I can't claim to be an authority. I have traveled in Latin America and thus do have some acquaintance with many of my subjects, although there are some of which I have no first-hand knowledge, for example, the water buffalo and yams. It is in my travels in Latin America that I have seen hungry people.

---

*This statement may be traced to Socrates, who is reported to have said, "Bad men live that they may eat and drink, whereas good men eat and drink that they may live."

The book has been written with the general reader in mind, and no particular background in biology should be necessary for understanding most of the topics. I had once assumed that the readers of a book such as this would have an elementary knowledge of human nutrition, but, judging from recent news releases, that assumption was unjustifiable, for malnutrition is not confined to the poor and uneducated but extends to the affluent and "educated" as well. Therefore, a brief treatment of nutrition is given (Chapter 3).

Although I have not tried to include all of my sources, there is a fairly extensive list of references given at the end of the book. This is included primarily for those readers who would like to pursue any subject in greater detail. Although I expect the same plants and animals to continue to serve as our principal foods for a long time to come, obviously the detailed knowledge concerning them will change as research makes more information available. Perhaps this is nowhere more true than in parts dealing with prehistory, for the next archaeological investigation may uncover new information regarding man's "invention" of agriculture and his earliest domesticated plants and animals.

The above paragraphs were written several months ago, and how true the last sentence has already proved to be. After the book had gone to press, a report was published of a new archaeological find of beans in Peru, dated at about 6000 BC, considerably earlier than any previous report. It was possible to add a footnote to the text concerning them. Two weeks later a paper appeared on "Earliest radiocarbon dates for domesticated animals" (Reiner Protsch and Rainer Berger, *Science*, vol. 179, pp. 235–239. 1973). The authors point out that for that for an accurate determination of age, conventional radiocarbon dates must be calibrated by tree-ring chronologies. From so doing it appears that some revision of previous dates for the time of the appearance of certain domesticated animals is in order. Of particular interest is that the new evidence indicates that southeastern Europe may have been as important as the Near East as a place of animal domestication.

Attention should also be called to a few other works that have appeared since the bibliography was completed. George Beadle has published an interesting semipopular account of the origin of maize (*Field Museum of Natural History Bulletin*, vol. 43, no. 10, pp. 2–11. 1972). Of interest in connection with my Chapter 10 is a paper by

Herbert Baker on human influences on plant evolution (*Economic Botany*, vol. 26, pp. 32–43. 1972). A book entitled *U.S. Nutrition Policies in the Seventies*, edited by Jean Mayer (W. H. Freeman and Company, San Francisco, 1973), should prove valuable reading to people who are particularly interested in Chapters 3 and 11.

There are many people to whom I am indebted for my interest in this subject and for help with the present book, but I shall single out only three. It was the late Edgar Anderson of the Missouri Botanical Garden who instilled in me the fascination of studying our cultivated species. My colleague, Paul Weatherwax, has taught me much of what I know about grasses as well as supplying several illustrations for this book. And to my former student, Barbara Pickersgill of the University of Reading, I owe thanks for many stimulating discussions concerning plants and man. I, alone, of course, take responsibility for what is said in the following pages. I should also like to thank those, particularly FAO, who have supplied illustrative material. Several of the drawings have been made by Joan Wood, a student at Indiana University.

January 1973   *Charles B. Heiser, Jr.*

1

The origin of agriculture

In the sweat of thy face thou shalt eat bread.
Genesis 3:19.

Man has been on earth for some two million years. Except for a minute fraction of that time he has been a hunter of animals and gatherer of plants, strictly dependent upon nature for food. He must, at many times during his long history as a hunter-gatherer, have enjoyed a full stomach when vegetable foods were abundant or ample game was available. Early man certainly must have experimented with nearly all of the plant resources, thus becoming an expert on which ones were good to eat. He became an excellent hunter and fisherman. Contrary to earlier opinion, recent studies suggest that he didn't always have to search continually just to find enough to eat and that at times he must have had considerable leisure time.

There undoubtedly were times and places, however, in which people did have to spend most of their waking hours searching for food, and hunger probably accompanied man throughout much of his preagricultural period. Certainly there could never have been much of an opportunity for large populations to have built up, even among the successful hunter-gatherers. Man probably lived in small groups, for with few exceptions a given area would provide only enough food for a few people. Disease and malnutrition probably contributed to keeping populations small; it is likely that there were also some sorts of intentional population control, such as infanticide.

Then about 10,000 years ago man's food-procuring habits began to change, and in the course of time man became a food producer rather than a hunter-gatherer. At first he had to supplement the food he produced with food he obtained by hunting and gathering, but gradually he became less dependent on wild food sources as his domesticated plants and animals were increased in number and improved. Through the cultivation of plants man became able to produce more food with less effort. Having a dependable source of food made it possible for larger numbers of people to live together. More mouths to feed were no longer disastrous but rather were advantageous, for with more bodies to till and reap, food could be produced more efficiently. As food production became more efficient, villages arose and in time cities came into existence, and civilization was on its way.

Along with food production, man found time to develop the arts and sciences. Some hunter-gatherers, as was already pointed out, must have had considerable leisure, but they never made any notable movement toward civilization. An important difference between hunter-gatherers and farmers is that the former are usually nomadic whereas the latter are sedentary. But even those preagricultural people, such as certain fishermen, who had fairly stationary living sites did not develop in civilizing ways comparable to the farmers. Agriculture probably required a far greater discipline than did any form of food collecting. Seeds had to be planted at certain seasons, some protection had to be given to the growing plants and animals, harvests had to be reaped, stored, and divided. Thus we might argue that it was neither leisure time nor a sedentary existence but the more rigorous demands associated with an agricultural way of life that led to great changes in man's culture. It has been suggested, for example, that writing may have come into existence because records were needed by agricultural administrators. Man was changing plants and animals to suit his needs; living in close relation with plants and animals was also changing man's way of life.

In recent years archaeological work has greatly increased our knowledge of the beginnings of agriculture, and without doubt future archaeological work will add a great deal more information. In contrast to previous generations of archaeologists who were mostly concerned with spectacular finds—tombs and temples, the contents of which would make showy museum exhibits—recent archaeologists have taken a greater interest in how man lived, what he ate, and how he managed his environment. A few charred seeds or broken bones may appear rather insignificant in a museum but they can reveal a great deal about early man's activities. As a result of recent work in archaeology, done in cooperation with scientists from many other fields, we are beginning to understand the ecology of prehistoric man in many different places of the world.

Our knowledge of what man ate and did thousands of years ago comes from the remains of plants and animals recovered from archaeological excavations. Unlike many of man's tools, which were made of stone and are indestructible, foods, being perishable, are only preserved where conditions are ideal. The best sites are in dry regions, often in caves, and from such sites we obtain remains to use in the reconstruction of the diet of ancient man. Other human artifacts, such as flint sickles and stone querns or grinding wheels, may provide clues about diet but leave us to speculate about what plants were being harvested and prepared, and whether these were wild or cultivated. Obviously the record of what ancient man ate is very incomplete and for many areas of the earth significant remains have yet to be found.

Drawings of animals, particularly from the later prehistoric periods, have come down to us and these sometimes (but not always, by any means) can be fairly readily identified, but it is animal bones, or even fragments of them, that provide the best clues about the animals that were in close connection with man. An expert zoologist can identify species from bones, but it is not always possible to say whether remains are from domestic or wild animals.

Plant remains comprise a variety of forms—most of them are seeds, but some fruit rinds, flower bracts, stalks, and leaves are found. A few remarkably well-preserved seeds are recovered, looking as if they had been harvested only a year before, but most are charred and broken. A skilled botanist can identify plant remains and it can be determined if they are from cultivated or wild plants.

Another source of information about what ancient man ate is coprolites—fossil feces. By suitable preparation they can be restored to an almost fresh condition (sometimes, it is said, including the

A

B

Figure 1-1
A. Archaeological dig at Coxcatlan cave, one of the sites in the Tehuacan Valley, Mexico, that has revealed early evidence of maize. (From D. S. Byers, ed., *The Prehistory of the Tehuacan Valley*, Vol. 1. Univ. of Texas Press, Austin, 1967. Used by permission.)
B. Increase in size of maize between c. 5000 BC and c. AD 1500 at Tehuacan. The oldest cob is slightly less than one inch long. (From D. S. Byers, ed., *The Prehistory of the Tehuacan Valley*, Vol. 1. Univ. of Texas Press, Austin, 1967. Used by permission.)

Figure 1-2
Carbonized emmer wheat from the archaeological site of Sesklo in Thessaly, Greece. (Courtesy of Jane Renfrew.)

odor). Whole seeds have been found in coprolites but most of the food material is highly fragmented and requires lengthy painstaking analysis for identification. Such analysis is highly significant because it tells us what was actually eaten, in what combinations, and whether it was cooked or raw.

Unfortunately, at times not all material collected at an archaeological dig is accurately identified, as has been shown for some of the early archaeological reports from Peru. Fortunately, however, the material recovered from archaeological sites is usually preserved in museums, and future investigators can examine the material to verify or correct identifications.

With the development of radiocarbon methods of dating it became possible to determine fairly accurately the time at which man began to cultivate plants. Sometimes radiocarbon dates, for one reason or another, may be open to suspicion, but when we have several dates from different material at the same site which show agreement, we have fair assurance that these are correct within a few hundred years.

The evidence that has accumulated over the past several years indicates that agriculture probably had its first origins in the Near East—although not, as earlier supposed, in the fertile river valleys of Mesopotamia, which were to be important centers of early civilization, but rather in semiarid mountainous areas nearby. Dates determined for flint sickles and grinding stones discovered there indicate that before 8000 BC man had likely become a collector of wild grain, and there is evidence that a thousand or so years later he actually was cultivating grains and keeping domesticated animals. Several sites are now known in the Near East (see Figure 1-3) that give evidence of early agriculture. One of the first sites to give such evidence was at Jarmo in Iraq where investigations were conducted under the direction of R. J. Braidwood. In deposits dated at 6750 BC, seeds of wheat and barley and bones of goats were found. At several other sites in the Near East dated at approximately the same time other evidence of cultivation is found. Since the plants in these sites apparently represent cultivated species, we must suppose that there was an earlier period of their incipient domestication, which may have lasted for a few hundred years or more. How long it takes a plant to become fully domesticated cannot be answered precisely and it probably varies considerably from species to species. In deposits accumulated after 6500 BC we find evidence of other plants being cultivated in the Near East and bones of various domesticated animals become more abundant.

Other centers of agriculture developed in the Old World. Whether these developments were stimulated from knowledge of agriculture in the Near East or whether they were independent developments is not certain, but the fact that some of them were based on completely different plants from those of the Near East might support the later view. For a long time southeastern Asia has been considered as an ancient center for domesticated plants but until recently there was no archaeological support for this view, which is not wholly unexpected since the climate for the most part in this area of the world is hardly conducive to preservation of food remains. In 1969, however, a report was published of an assemblage of plants from Thailand, including possibly a pea and bean, dated at 7000 BC. As it is not definitely clear whether the plants recovered represented wild or cultivated species, we cannot yet say that agriculture was practiced as early here as it was in the Near East. We do not yet know when rice, which was to become the basic food plant of southeastern Asia, was first brought under cultivation.

In the New World, agriculture began a few thousand years later than in the Near East and had its origins in Mexico. Through a series of excavations directed by R. S. MacNeish, we now have a remarkable sequence of plants giving evidence of the period of incipient

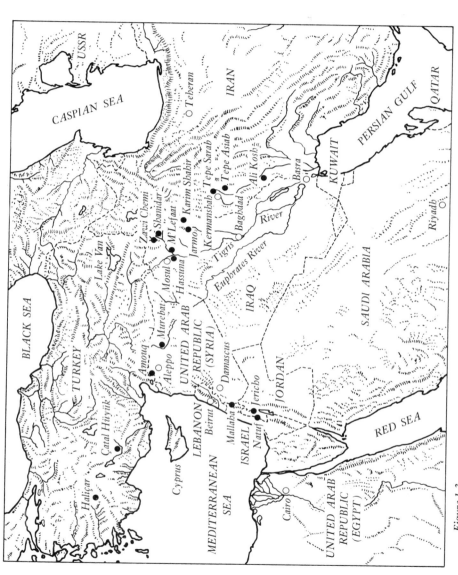

Figure 1-3
Selected archaeological sites (solid dots) that show evidence of early agriculture in the Near East.

domestication. Some indication of the earliest cultivated plants is found in the mesquite-desert regions of southwestern Tamaulipas with gourds, squashes, beans, and chili peppers being found at levels dated as between 7000 and 5500 BC.

Following investigations at Tamaulipas, MacNeish made deliberate efforts to search for evidence of the domestication of maize, which eventually was to become the most important plant in the Americas. A group of caves in arid highlands near Tehuacan in south central Mexico showed promise, and a series of excavations was begun in 1961. The results give us the best picture yet of the transitional stages leading to full-scale agriculture. Man was probably in the Tehuacan area by 10,000 BC and for several millenia he depended on wild food sources, both plant and animal. Gradually more and more plants were cultivated, some perhaps having been domesticated at this site, others introduced from other regions. The first suggestion of cultivated plants occurs in material dated at about 5000 BC, with maize, squash, chili pepper, avocado, and amaranth being found. These plants were definitely cultivated during the next period (4900–3500 BC), together with, toward the end of the period, various fruits and beans. During the next thousand years other plants were added, including cotton and two new kinds of beans. The dog, which from historic records is known to have been an important food item in Mexico, is first associated with man in the archaeological record at this time. We find that at about the beginning of the Christian era, the inhabitants of Tehuacan had also acquired the turkey. From remains of the same period, there are reports of some other plants: guava, pineapple, and peanut. The presence of these plants would be of particular interest, for the peanut is definitely South American in origin and the pineapple and guava perhaps are also, which would suggest that the peoples of this area had somehow established contact with South America. None of these plants has been found in any other archaeological sites in Mexico to date. A study of historical records of both peanuts and pineapple would suggest that they are recent arrivals in Mexico, perhaps after the coming of the Spanish.

A second important center for early agriculture in the Americas was in Peru, where its origin was somewhat later than in Mexico. Considerable plant material has been recovered from archaeological sites in the dry coastal areas but as yet we do not have any sequences of plant materials that compare with that from Tehuacan. One of the

Figure 1-4
New World archaeological sites (solid dots)
mentioned in text.

important sites is Huaca Prieta in northwestern coastal Peru, which
was investigated by Junius Bird. From this and other deposits we
know that gourds, squashes, cotton, lima beans, and chili peppers were
some of the first plants cultivated. As yet there is no clear indication
that agriculture was practiced much earlier than 3000 BC.

A cursory examination of the first plants cultivated in Peru and the
fact that agriculture appeared there much later than in Mexico might
suggest that agriculture came to Peru from Mexico. But a critical
examination reveals that the beans, chili pepper, and cotton belong to

species different from those cultivated in Mexico at the time and probably are derived from wild species indigenous to Peru. The picture is not as clear for the squashes and gourds but these too may prove to be domesticates of wild plants from South America rather than imports from Mexico. Thus while we can't rule out the possibility that the idea of agriculture came from Mexico, it certainly appears that local species were brought under domestication in coastal South America before domesticated plants arrived there from Mexico. The earliest evidence of maize in coastal Peru is dated at about 2000 BC. As yet we have little information on the beginnings of agriculture in highland Peru but some recent evidence indicates that it was earlier than on the coast. A recent archaeological investigation found evidence of maize in the highlands at the time that the first cultivated plants appear on the coast.* Since maize had its origin in Mexico, its presence in South America would indicate the establishments of contacts with Central America or Mexico.

From the foregoing account it can be seen that agriculture arose in widely separated parts of the earth, probably quite independently from place to place, although there exists a possibility that the knowledge of agriculture reached Peru from Mexico. Could the idea of agriculture also have come to the New World from the Old? The New World was peopled by immigration across the Bering Strait long before agriculture was known. If there were subsequent crossings at this place, it was by men who were hunters rather than agriculturists. As agriculture began in the Old World more than a thousand years earlier than it did in the New, we would have to postulate a long ocean voyage at a very early date to account for the knowledge being brought to the New World. Some anthropologists have postulated that there were such voyages in prehistoric times, but much later than the time at which agriculture was established in Mexico. Therefore, it seems highly unlikely that agriculture had but

---

*Since the above was written, a report on archaeological plant discoveries from an intermontane Peruvian valley has appeared (L. Kaplan, T. F. Lynch, and C. E. Smith, Jr., *Science, 179:* 76–77. 1973.) Both domesticated common beans and lima beans are found in deposits dated at about 6000 BC. This find would indicate that not only is agriculture earlier in the Andes than on the coast, but also that agriculture appears to have had an origin in Peru as early as, or perhaps even earlier than, in Mexico. This report also lends support to the hypothesis that agriculture had completely independent origins in the two places. Although common beans now appear at early archaeological levels in both Peru and Mexico, this could be explained by separate domestications of the same kind of wild bean in the two areas.

Figure 1-5
Mexican girl grinding maize. (Reprinted with permission of The Macmillan Company from Paul Weatherwax, *Indian Corn in Old America.* Copyright © 1954 by The Macmillan Company.)

a single origin. It is, in fact, likely that it had several origins in both the Old and the New World.

An examination of the list of food plants from all the early sites in both the New and the Old World reveals that all of the plants were propagated by seed. A large number of man's food plants, including such important ones as both the white and the sweet potato, manioc, yams, bananas, and sugar cane, are propagated vegetatively—by stem cuttings, tubers, or roots—rather than from seed. Some people, notably the geographer Carl Sauer, have reasoned that man's cultivation of plants probably began with vegetative propagation, arguing that such cultivation is much simpler than seed planting. The archaeological record, unfortunately, has not been able to provide us with clear-cut answers, for many of the vegetatively cultivated plants are crops of the wet tropics, areas where the preservation of prehistoric food materials is rather unlikely. Moreover, even in dry areas, tubers and other fleshy plant parts are far less likely to be preserved than are relatively dry material, such as seeds. While we cannot, perhaps, entirely rule out the possibility that agriculture based on vegetative propagation was earlier than seed-propagation agriculture, it seems fairly clear that it was seed planting that led to the most profound changes in the life of man. All the early high civilizations whose diets are known to us were based on seed-reproduced plants—wheat, maize, or rice—some with, and others without, accompanying animal husbandry.

Following the domestication of plants and animals, the next great advance in agriculture came with the control of water. Irrigation arose in the Near East around 5000 BC and in Mexico shortly after 1000 BC. With irrigation man could produce considerably more food in many areas; as a result a few people could produce enough food to feed a large population, permitting other people to spend their time in pursuits of the arts and crafts and of religion. Elaborate temples, many of which are standing today, constructed by the early societies that had perfected methods of irrigating their crops, testify to the amount of human labor that was made available for other pursuits.

Another important development in Old World agriculture was the use of animals to prepare the fields for planting, which was never done in the New World in prehistoric times.

With the domestication of plants and animals, man should have had a dependable food supply and, so it might be thought, hunger should have disappeared from the earth, but as any intelligent person is acutely aware it is still very much with us today. Man, in fact, still needs to learn to live with nature. With the advent of agriculture man began changing his environment drastically. Irrigation, which initially led to greater food production, eventually destroyed some of the most fertile areas. Without adequate drainage, irrigation leads to an accumulation of salts in the soil that few plants can tolerate. That this happened in prehistoric time in the Near East is evident from archaeological findings; for barley, which is more salt tolerant than wheat, replaced the latter plant in some regions after irrigation was developed. The use of animals to till the soil led to increased areas being planted, which in time must have been accompanied by increased soil erosion. Then along with the plants and animals that man brought under his control came others that he did not want and could not control. Rusts, smuts, and weeds soon found man's plants and fields a fertile territory for their development, and insects, rodents, and birds moved in to appropriate some of man's new foods for themselves. Competition among men for the more fertile agricultural land led to warfare on an escalating scale, for which the powers of some of the domesticated animals were used. Hunger has always accompanied war.

A

B

C

Figure 1-6
A. Hoeing and horse-drawn plow, reproduced from cave drawings in northern Italy. (Redrawn from a photograph by Emmanuel Anati.)
B. Hoeing and ox-drawn plow, from decorations on a tomb at Beni Hasan, Egypt. (Courtesy of Egypt Exploration Society.)
C. Present-day plowing with oxen in Egypt. (Courtesy of FAO.)

Deserts now occupy many of the areas where high civilizations once flourished. Natural climatic change may in part be responsible for some of these deserts but man likely contributed by his misuse of soil and water. Man's alteration of the environment, which began in a modest way 10,000 years ago, continues to the present on a scale never known before.

2

Seeds, sex, and sacrifice

O goddess Earth, O all-enduring wide expanses!
Salutation to thee.
Now I am going to begin cultivation.
Be pleased, O virtuous one.

Ancient Sanskrit Text

The work of the archaeologist has revealed a great deal concerning the "invention" of agriculture. We now have some idea about where and when it occurred and what plants and animals were involved, but we do not know why man domesticated plants and animals. The answer may be very simple: Man may have wanted a dependable source of food that was close at hand, and what would be more obvious than to bring animals into confinement and to grow plants in some suitable place near the home. In James Michener's best-selling novel of a few years ago, *The Source*, we read that thousands of years ago the wife of Ur transplanted wild grain near her dwelling; and although her first efforts failed, eventually she succeeded in bringing it under cultivation. How she hit upon the scheme of collecting seeds and planting them, rather than transplanting already established plants, is not revealed to us. This fictional account is of interest, nevertheless, because it serves to illustrate the "genius" theory of the origin of agriculture, which would have agriculture arise through the efforts of a single brilliant person. Most archaeologists have been unwilling to accept such a hypothesis, perhaps because it explains nothing of the circumstances leading to cultivation. This is not to say that man was any less intelligent ten thousand years ago than he is today, and the observation that a seed germinates to give rise to another plant of the same kind was probably well known to those who depended on seeds for their main source of food. Knowledge of seed germination would not necessarily have led, however, to the planting of seeds. As Kent Flannery has pointed out, "a very basic problem in human culture is why cultures change their mode of subsistence at all." Our problem is to explain why man changed from a hunter-gatherer to a farmer. The archaeological record does not provide any definite answers.

Archaeologists have speculated upon this subject. V. Gordon Childe, who gave us the term the Neolithic Revolution for man's invention of agriculture, believed that a climatic change resulting in desiccation brought man and animals together where there was water, and through this association the domestication of animals was stimulated. R. J. Braidwood postulated that without major climatic change food production developed "as the culmination of ever increasing cultural differentiation and specialization of human communities." He assumes that it was human nature to invent agriculture when man had reached a great familiarity with his plant and animal resources. Lewis R. Binford, rejecting both Childe's and Braidwood's arguments, felt that demographic pressure was instrumental: An increase in population density led man to attempt to manipulate the environment in order to increase food production. Kent Flannery, who has had the advantage of working at sites in both the Old and New World where agriculture had its origins, is inclined to accept certain of Binford's reasoning and suggests that people at the margins of the optimum wild food zone in the Near East initiated cultivation in an attempt to duplicate the dense stands of cereals found in the optimum zone. For Mexico he has suggested that genetic improvement in early maize produced a "positive feedback" leading to the attempt to introduce maize into new environments. The geographer Carl Sauer has favored the idea that cultivation of plants first arose among fishermen. Such people would have been more-or-less sedentary and would have had a dependable source of food, giving them the time and stability to experiment with cultivation. Some botanists have subscribed to a "dump-heap" origin of agriculture, popularized by Edgar Anderson.

In the rubbish heaps of man's campsites, seeds might have been dropped and parts of plants might have been discarded. Such refuse heaps, being rich in nitrogen, could have given rise to vigorous plants that were in turn used by man.

A hypothesis at variance with all the previous ones has recently been proposed by Jane Jacobs in *The Economy of Cities*. She believes that cities gave rise to agriculture and not the reverse as is generally held. The early cities, she postulates, arose as trading centers and agriculture actually developed in them and was later moved to outlying areas. She presents an interesting case, particularly about how animals brought to the cities for barter would have been kept alive until needed, which obviously might be a first step toward their domestication. She uses Çatal Hüyük, a city known to have been in existence in Turkey at 6000 BC, as evidence in support of her ideas. While it is becoming increasingly evident that some of the early cities may have originated as trading centers, the archaeological evidence available so far indicates that their basic foods were from domesticated sources presumably in nonurban areas. Neither Jacobs' hypothesis, nor any of the others except Flannery's, provides good explanations of why early man might have begun to plant seeds.

There are many other questions concerning the origin of agriculture that have no ready answers. For example, why didn't agriculture begin earlier? After all, man had been around for more than a million years before he domesticated plants and animals. It has to be admitted, however, that we can't be absolutely certain that domestication had not been tried earlier. The archaeological evidence is never complete. Perhaps there was one or even many attempts at domestication before it was successful. Certainly the beginnings were feeble and halting and success was far from assured. If cultivation had been tried and the crops had failed or been lost through some climatic catastrophe, the people might have given up all attempts, perhaps because of their fear of having made the gods angry, and gone back to a hunting-gathering economy. It could be that Braidwood's hypothesis is correct—that man was not ready for agriculture until his culture had reached a certain level. This is not a very precise answer, for certainly few tools were needed at the start. The plants and animals that later became man's principal foods were available for domestication before 10,000 BC. We can hardly suppose that man didn't have a good knowledge of them. Could it be that the concentration or scarcity of plants and animals in a given region had some bearing? This brings us to ask if there was some environmental factor that played a decisive role in the origin of agriculture.

For some time it was thought that there had been a climatic change leading to increasing aridity in the Near East before man practiced agriculture, and as was pointed out above, Childe used this in an attempt to explain the origin of agriculture. There was no evidence for such a change, however, and others maintained that the climate in the Near East had been relatively stable since long before agriculture began. Recently, however, evidence has been supplied that there was some change in climate in the Near East, although not of the nature that Childe had supposed.

In 1968 an analysis of pollen deposits in two lake beds in Iran by H. E. Wright, Jr., indicated that there was a shift in climate about 11,000 years ago. Through the analysis of fossil pollen it is possible to determine what plants grew in a given area in former times and from this the climatic conditions under which the plants grew may be inferred. The pollen analysis in Iran indicated a shift from a cover of herbaceous plants to one in which oak and pistachio trees predominated, which is interpreted as a change from a cool steppe to a warmer, and perhaps moister, savanna. Such a change would obviously have an effect both on the plants and animals present and consequently on man's eating and food-procuring habits. There are no comparable pollen remains from the sites of early agriculture in the Americas, but Kent Flannery's analysis of the animal remains at Tehuacan indicates the presence of several animals before 7000 BC that were not found at later times, which might be evidence of a comparable climatic change. The disappearance of these animals might be interpreted as resulting from a change in climate, which need not have been as drastic nor in the same direction as the climatic change in the Near East. Again, there would have been a gradual change in man's food resources as a result.

How else could the environment have figured? In both the Near East and Mexico the earliest known sites of agriculture are from semiarid, somewhat hilly or mountainous country. Is this simply owing to a natural bias for preservation in such settings? Evidence of agriculture is certainly far more likely to be preserved in arid or semiarid regions than in more mesic areas. Or is it possible that such places did favor the development of agriculture? Although this sort of environment may not seem ideal, there would have been certain advantages. These areas of diversified terrain would have provided a number of microenvironments appropriate for different species of plants and animals, which would, in turn, have offered man a considerable array of wild foods and also of potential domesticates. Only limited travel would have been necessary to secure enough to eat, and thus man could have been sedentary, at least at certain times of the year, a necessity for the establishment of cultivation. Low rainfall, as long as it was adequate for plants at certain seasons, may have offered many advantages to early agriculturalists in that there would have been no heavy plant cover to remove previous to planting, and there would likely have been fewer weeds and various plant pests, both insect and fungal, than in better watered areas.

All of this brings us back to the question about why man would change his mode of subsistence at all. If man had an abundance of wild food sources, why did he ever bother to domesticate plants and animals? We don't know for sure that the people who gave us agriculture were in such a position, but the possibility exists that there was an abundance of wild wheat and barley waiting to be gathered in the Near East where agriculture had its origin. On the other hand, if man had a limited food supply would he have ever had an opportunity to start domestication? If he had to spend most of his time moving around to find enough to eat, he would never have been in one place long enough to give attention to domesticating plants and animals. Necessity, as Carl Sauer has written, would not be the mother of invention in the case of agriculture.

To get out of this dilemma we could postulate that there was an intermediate situation with a nearly adequate food supply but meager enough that man would have been stimulated to search for new methods in order to procure more. Perhaps there were people living close enough to abundant stands of wild grains to know about them, as Flannery has suggested, who looked for ways to get such growths near their abodes. We might assume that such people would not have been in such dire need of food that they had to spend all of their time looking for it, but that they would have had some leisure time for experimentation and a somewhat sedentary life. Certainly, as previously pointed out, the people who originated agriculture must have been well acquainted with their animal and plant resources and probably had observed long before that a seed would give rise to another plant like the one from which it came. But someone had to make a practical application of this knowledge. Did this happen intentionally or accidentally? The answer is not known, nor do we know whether one person in a given area had the idea, which spread as a result of his success with it, or if a number of people in a given area might have hit upon the idea independently of each other at about the same time. The archaeological record does not rule out the latter possibility.

Thus far we have been considering that the origin of agriculture was intentional. Could it have had its origin by accident, that is to say, might agriculture have been originated, not in a direct attempt to secure more food, but as a by-product of some other activity of man? Such origins were postulated during the last century, but in order to appreciate them it becomes necessary to make an excursion into primitive man's beliefs concerning himself and his environment.

Our knowledge of man's early religion is, of course, very fragmentary and interpretations of the religious significance of material remains that have come down to us are highly speculative. It seems quite evident, however, that birth, death, and food were of fundamental importance to primitive man and that his concern with each of these affected most of his activities. There were many mysteries connected with all of these, and they became interrelated. In Paleolithic times, the animals, although they were killed for food, were nevertheless considered akin to man, and the seasonal cycles of the death and rebirth of vegetation were obviously thought to be related to man's own life. A worship of trees or plants was probably an early manifestation of man's religion and was carried down through Neolithic times to become an important part of formal religion in early historical times. Such beliefs still survive in the folklore of many people. In trees and other vegetation man recognized a life-giving power, related to his own and that of animals. His own fertility, that of animals, and that of vegetation were obviously closely related.

Associated with human burials from early times in many parts of Europe and Asia are small female figurines or Venuses, made of bone, stone, or ivory. The face may be entirely lacking or crudely represented and the sexual features are exaggerated. The well-developed abdomen on some figures is thought to represent pregnancy; others with a prominent vulva perhaps indicate women giving birth. The rather extensive distribution of such figurines has been interpreted as evidence of a widespread mother-goddess fertility cult in Neolithic times; eventually various distinct goddesses were worshipped as regional religions developed.

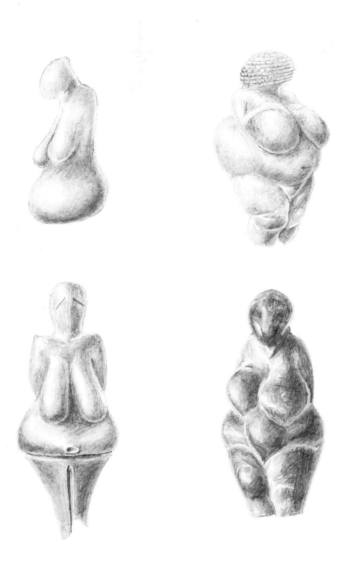

Figure 2-1
Mother-goddess figurines from the Near East and Europe.

Whitworth College Library
Spokane, Washington

Recognition of the sexual significance of the male* apparently increased during later Neolithic times, for phallic symbols from this period have been found in various regions by archaeologists. Eventually a young god—in some places considered to be a brother or son of the goddess—became known in primitive religion. The union of the goddess with this god was then regarded as being responsible for fertility. Among some human groups the living king was thought to be a god and as such he was an important figure in sacramental marriage ceremonies relating to fertility. From this it is not difficult to see that all human sexual intercourse became symbolically associated with fertility. In later times human sexual intercourse became a part of festivals held in the fields at the time of planting in order to promote growth of the crops and of other festivals associated with agriculture. The plow itself seems to have been first designed as a phallic symbol, representing man's role in bringing fertility to mother earth. Sexual offenses were thought to impair fertility. Thus we see that in early times sex was considered to be sacred among some people.

With the development of agriculture, other gods joined the mother goddess and her consort to share their duties; the earth, the sky, the rain, and other natural elements were among the special domains of the various gods. In many cultures the principal male god of the pantheon (Zeus is one of the best known), who had been born of the mother goddess, eventually assumed a dominant role. The relationships among the gods became complicated and their responsibilities less clear-cut, but there were always some whose main concern was with fertility, being remnants from the earlier fertility cults. Fertility ceremonies continued during the reign of the Roman Empire and the excesses committed at some of the festivals led the Romans to decree laws against their observances. Such ceremonies are still widely practiced among "primitive" people in many parts of the world today, and sometimes by "advanced" people as well, although not always in

---

*That the female had a role in fertility, of course, could never be doubted. Exactly when the full significance of the male was recognized, is, of course, not known. Some present-day primitives do not understand the male's contribution to procreation, thinking rather that intercourse may be necessary to allow a spirit to enter the womb or to make it easier for childbirth. This idea contrasts strongly with that prevalent among Western peoples during the late Middle Ages, when it was thought that the sperm contained the fully made child (the *homunculus*) and the female simply served as a house for its early development.

a form recognized as such. Most people participating in May Day celebrations today are probably unaware that its original purpose was to promote the well being of vegetation.

The temple prostitutes mentioned in the Bible were participants in rituals derived from the earlier sacramental marriages. The prostitutes and the fertility festivals were denounced by the Israelites, who originally were pastoral desert-dwellers and not tillers of the fields. As their religion was monotheistic and their god, Jahweh (or Jehovah), was not a fertility deity, the ceremonies could not be viewed by them as legitimate religious activity nor the women as anything but harlots. It has recently been pointed out that many of our present-day environmental problems may stem from the Judeo-Christian concept that the earth and everything on it were put here solely for man's use. Would things perhaps have been different if the religions of the developed nations had been derived from the fertility cults with emphasis on reverence for mother earth and her creatures, both plant and animal?

How early sacrifice developed is not known, but there is evidence that it was practiced at Tehuacan before 5000 BC and it became well established in most early agricultural cultures. Some people have postulated that humans were the first victims and were later replaced by animals, but it is perhaps just as likely that humans replaced animals as cultures reached a certain stage of advancement. Practically all of the domesticated animals have been used in sacrifice at one time or another, sheep, goats, and cattle all being prominent, as readers of the Old Testament are aware. No animals were more important than cattle. A cattle cult apparently was well established at Çatal Hüyük in Turkey by about 6000 BC. As part of fertility cults, cattle became associated with the gods themselves and became prominent figures in many primitive religions.

Various reasons have been advanced to explain sacrifice, the simplest being that it was to honor or appease the gods. It is likely that it was much more complicated than that, and probably sacrifice had a dual role. The human or animal being sacrificed represented the grain or the produce of the field and at the same time the people who were to partake of it. The sacrifice would bring about a desanctification of the spirit of the plant to make it safe for humans to eat and also would assure a future bountiful harvest of the fields. The victim in some cultures was the king. As this didn't prove to be popular with kings, in later times a lesser person was substituted. The dying king, or his representative, symbolized the dying vegetation and was replaced by

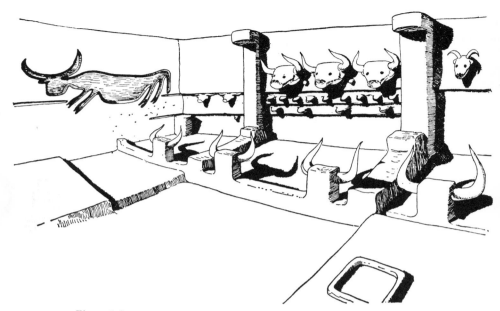

Figure 2-2
Restored cattle shrine at Çatal Hüyük (c. 6000 BC). From James Mellaart, *Çatal Hüyük: A Neolithic Town in Anatolia.* Thames and Hudson, Limited, London, 1967. (Used by permission.)

a new king to represent the resurrection of vegetation and life itself. After the abandonment of human sacrifice, effigies were sometimes used to serve the same purpose, a practice that survived until recent times in some places. That human sacrifice may not have completely disappeared, however, is evident from an account from Tanzania. In 1959 several farmers of the Wangi tribe were arrested for violation of a witchcraft ordinance. The farmers were suspected of *Wanyambuda*, an ancient tribal fertility rite in which fields were sprinkled with "medicine" made of seeds, blood, and parts of human bodies.

Although this account of primitive religion is oversimplified and very incomplete, there can be little doubt that man's early religion was intimately involved with his quest for food. As this is true, we might inquire if in man's religious beliefs we can find some clues about how domestication began.

The dog has been thought to be man's first domesticated animal; dogs have served as food for man in historic times and doubtless did

Figure 2-3
Inca sacrificing a llama.
(From Poma de Ayala, c. AD 1600.)

in prehistoric times as well. In some cultures the dog may have been more important as man's hunting companion than as food, and the suggestion has been made that the dog furnished great aid in bringing other animals under domestication, just as it continues to serve pastoral people today. It seems unlikely that a religious motive could have been involved in the domestication of the dog, a fuller discussion of which will be postponed until the next chapter.

It is also entirely possible that religion had nothing whatsoever to do with the domestication of other animals. The presence of dogs may have inspired attempts at other domestications. Perhaps men brought home young animals whose mothers had been killed. The

Figure 2-4
Cattle. (Courtesy of FAO.)

young animals may have been nursed by the women and became pets, leading to their domestication. This hypothesis has been considered the most likely by many authorities on the subject.

The idea that religion may have been involved, however, is not a new one. Eduard Hahn, a German geographer, who published just before the turn of the century, maintained that cattle were domesticated in order to secure animals for sacrifice in connection with lunar ceremonies to insure fertility. Hahn believed that cattle were chosen for sacrifice because their crescent-shaped horns resemble the new moon. Other reasons could be suggested. The wild bull was a ferocious animal and would have inspired both fear and admiration. Perhaps man sacrificed young male animals to secure their strength and virility. That cattle became highly preferred for sacrifice and eventually even became sacred in certain cultures is quite evident, but this, of course, in itself does not necessarily mean that cattle were domesticated for religious rather than utilitarian motives.

Recent archaeological discoveries indicate that cattle were not the first ruminant animal to be domesticated, both goats and sheep being earlier. Are we to suggest that perhaps both of these animals were domesticated for religious rather than economic reasons? Certainly both sheep and goats were widely used for sacrifice, and there is an old Sumerian incantation referring to a wild goat being used in sacrifice. If wild animals were used for sacrifice, man would not only have to capture them alive but would then have to keep them until the appropriate time. This could have been the first step in domestication. It is, of course, not necessary to try to account for the domestication of all the animals through religious considerations, for once one animal had been domesticated certainly the idea to domesticate others might have occurred to the same people or others who learned of it.

Although Hahn's thesis for the origin of animal domestication has received serious consideration by some modern authorities, the work of Hahn's contemporary, Grant Allen, in regard to plant domestication has been largely forgotten. Allen, a Canadian novelist, science writer, and philosopher, in a paper published in 1897 speculated upon the origin of seed cultivation. Reasoning that knowledge of seed germination and the clearing of the land were essential for sowing to be a success, he looked for ways in which man might have stumbled upon these requirements. He wrote:

> I believe there *is* one way, and one way only, in which primitive man was at all likely to become familiar with these facts, but it is one so startling, and at first sight so seemingly improbable, that I hesitate to suggest it save in the most tentative manner. Yet I shall try to show that all the operations of primitive agriculture very forcibly point to this strange and almost magical origin of cultivation; that all savage agriculture retains to the last many traces of its origin; and that the sowing of the seed itself is hardly considered so important and essential a part of the complex process as certain purely superstitious and blood-thirsty practices that long accompany it. In one word, not to keep the reader in doubt any longer, I am inclined to believe that cultivation and the sowing of seeds for crops had their beginning as an adjunct of the primitive burial system.

Primitive men buried tools and food with the dead for their use in the next world. This act, according to Allen, would entail the unwitting sowing of seeds. The annuals that were man's main food plants would have thrived in the cleared disturbed soil in the vicinity of a grave.

> What conclusion would at once be forced upon him? That seeds planted in freshly turned and richly manured soil produce threefold and fourfold? Nothing of the sort. He knows nought of seeds and manures and soils; he would at once conclude, after his kind, that the dreaded and powerful ghost in the barrow, pleased with the gifts of meat and seeds offered to him, had repaid those gifts in kind by returning grain for grain a hundred-fold out of his own body.

Man would have then observed that the plants grew exceptionally well only in newly made graves, and would have concluded that a new burial would have to be made every year. This would lead to the necessity for human sacrifice—a magical act to secure more plants. He would note that plants grew well not only on the grave but in the disturbed area nearby, and thus the area of turned soil would have been increased.

To support his contention, Allen brings together many examples from all over the world of human sacrifice in connection with agriculture, and, as he correctly points out, many other examples could be given. In many such examples, however, we find that the body is not buried, but cut and scattered along with the blood over the field.

Allen's hypothesis offers a possible explanation of why man learned to turn the soil for planting. The disposal of the dead has taken various forms among primitive people, but burials, as Allen supposes, were widely practiced. Although many burials were in caves or under houses and hence would not have been conducive to plant growth, interment in open fields was probably common. Burials in temples probably came much later than the origin of agriculture, and were, moreover, probably reserved for royalty or priests.

The source of the seeds, however, poses more of a problem unless they came from wild plants growing near the grave site. From the earliest times a dead man's belongings and other objects have been buried with him; but, if food were placed with a corpse, would the seeds be uncooked and hence capable of germination? If the seeds were not cooked would they be placed sufficiently near the surface of the ground to germinate?

Figure 2-5
A. Prize cattle decorated for a festive occasion, India. (Courtesy of FAO.)
B. "Demon" guarding field of modern rice, India. Improved strains of plants and animals exist side by side with traditions and ceremonies in many parts of the world. (Courtesy of Rockefeller Foundation.)

Another possible objection to Allen's hypothesis relates to the sacrifice itself. Although there can be no doubt that sacrifice often accompanied planting, we have no knowledge whether this was practiced at the beginning of agriculture. In fact, the earliest sacrifices may have been of animals and the carcasses likely would not have been buried. For that matter could Allen be incorrect in regard to man's conclusion about the necessity of a dead body? Perhaps someone did actually conclude that cleared disturbed soil had a direct relation to plant growth. Thus we see there are some difficulties to the acceptance of Allen's hypothesis, at least in its entirety.

Are there ways in which man's religion might have influenced seed planting other than by connection with human sacrifice? Primitive man must have had many rituals and ceremonies associated with both planting and harvest; and while, of course, we have no direct knowledge of these, we can make conjectures from the knowledge we have from early historical times and from primitive cultures still surviving. Prominent among these are ceremonies devoted to the "first fruits" or the "last sheaf" of the harvest. We may conclude that the early seed collectors recognized spirits in the plants. In fact, we know from the early historical record that the harvest has not always been a joyous occasion, as might be supposed, but was formerly accompanied by much sadness and lamentation as the body of the "spirit" of the grain was reaped. As a propitiation to the spirit, man might have returned a token offering of the seeds collected, either the "first fruits" or "last sheaves," to the spirit. This offering could have served the same purposes as a sacrifice in removing a taboo from the plant to make it safe for mortals to eat and at the same time making assurance of a rich growth of the grains in the following year.

The seed offering might have been scattered over the field from which it had been gathered, and examples are known of people who regarded the last sheaf as sacred and saved it for scattering over the field along with their seed in the next season. Perhaps the seed offering might actually have been buried in the soil with the recognition that mother earth was the source of life. Certain Arabs are actually known to have buried a seed offering and marked the grave with stones.

Thus we might postulate that the first seed planting was a magico-religious act to appease the gods. Such rituals, we would have to assume, took place among preagricultural seed collectors and we can imagine the next steps. In some place man would have recognized that the sacred sowing yielded plants. These would have been harvested in the next season and some of the seeds would have again been returned to the gods via the soil. In time ceremonial plantings would become larger and larger, and intentional cultivation would be on its way.

The "first fruits" hypothesis for the origin of seed planting also offers some difficulties. First, there is no adequate explanation in it about why man learned to till the soil. Also we have to assume that for some reason the planting was eventually transferred from the harvest season to the spring. Other objections, of course, could be raised. On the other hand, the "first fruits" hypothesis offers a possible explanation for the rapid improvement of cultivated plants after planting was initiated. We might ask why would man save his best seeds for planting rather than eating them? Obviously, if the seeds were for the gods, they would have been the largest, and most nearly perfect, or perhaps from plants showing unusual characteristics. We might postulate that artificial selection began to operate with the first offering of seeds to the spirits of the plants.

Although it seems too far-fetched to deserve much consideration, another possible explanation of the beginning of seed cultivation might be entertained: Could the origin of the planting of seeds some-how be associated with human reproduction? If the earth, as a mani-festation of the mother goddess, was regarded as the womb for vegetation, perhaps there was some concept that seed would have to be planted in her just as man planted his "seed" in women. This would mean that man would have to have had some concept of his own role in reproduction at an early date. From the art work on certain early Babylonian monuments, we know that the people in parts of the Near East placed clusters of flowers from male date trees in the female trees in order to secure better fruit set. Although the representations were made long after the origin of agriculture, such a sophisticated knowledge must have had much earlier antecedents. This early recognition of the role of male flowers is all the more amaz-ing when we realize that sex in plants was not "discovered" until the end of the seventeenth century and the idea was not generally ac-cepted until much later. It would hardly seem unreasonable that some people did appreciate the significance of the male in human repro-duction 10,000 years ago. The historical record is too late to be of any help to us, but it may be significant that in Sumerian, the earliest written language, the word *numum* was used for both seeds of plants and the "seeds" of animals, and we also find that later the Greeks used the same word for both seed and human sperm.*

Once man had acquired the knowledge of how to plant seeds—by whatever means or combination of means—of one plant, we might suppose that other plants soon followed it into cultivation. The idea of planting would have diffused and still other plants would have been brought under cultivation in other areas. However, as was

---

*We know today, of course, that a seed and a sperm are not at all equivalent. A seed develops from an ovule after there has been a union of a sperm with the egg contained within the ovule. The ovules, sometimes but not correctly called "immature seeds" or "unfertilized seeds," are found in the ovary of the flower of the higher seed plants.

Figure 2-6
Winged guardian spirit pollinating flowers of the date palm. From an Assyrian
bas-relief of the ninth century BC. Nimrud, Iraq. (Boston Museum of Fine Arts.)

pointed out in the first chapter, it seems likely that agriculture had several independent origins. Are we then to account for all origins through some sort of magico-religious rites? Not necessarily, for there may have been different avenues to the origin of planting; for example, we can not rule out intentional efforts on the part of man after some sort of "happy accident," but it may not be unreasonable to assume that magic or religion often played a role. Rites and ceremonies similar to those that were found in the cradle of agriculture of the Old World were found in most other parts of the world as well. They are, in fact, still widely practiced in many parts of the tropical world. Modern science has still not wiped out all such beliefs, which we sometimes call superstitions, within the developed nations. For example, some farmers in the United States still believe that planting should be done only during certain phases of the moon; others believe that if a menstruating woman walks through a garden the crops will fail.

The explanations of why and how man "invented" agriculture, clearly, are far from certain. Many questions remain concerning the development of agriculture. Did man domesticate plants first or animals? Why did man select certain plants and animals for domestication? What happens to a plant or animal in the process of domestication? Clear-cut answers aren't readily available for all of these questions, but some of them will be answered in the chapters that follow, in which we shall examine man's principal domesticated species.

3

We are what we eat

If you do not supply nourishment equal to the nourishment departed, life will fail in vigor; and if you take away this nourishment, life is utterly destroyed.

LEONARDO DA VINCI

Man is, of course, omnivorous. Some people have held that before agriculture was well developed, meat and fish were man's principal food. It is not certain that this was so in his earliest period, for monkeys and apes are primarily vegetarian, and very early man may well have been too. However, the association of broken animal bones with some of the remains of man's forebears suggests that the meat-eating habit was acquired very early.

Man's omnivorous character helps explain how he acquired such a wide distribution over the earth's surface. He could find suitable foods almost everywhere he went. Although as a species man eats just about everything, any particular human community selects certain plants and animals for consumption. This is certainly true today and probably extends far back into man's prehistoric period. We can surmise that early man experimented at times with all of the possible food resources of his environment,* but some became preferred over others. We know little of why he made the choices he did, but it has been suggested that palatability—such things as taste, texture, odor, and color—played an important role. But did this assure him of all the necessary nutrients? John Yudkin answers in the affirmative. "When he ate what he liked, he ate what he needed." Obviously had he not eaten what he needed he would have been eliminated by natural selection.

Use of fire was one of the early important cultural traits acquired by man. Fire not only gave him a source of warmth, a means of protection from wild animals, and a tool to assist him in capturing animals, but a new way to prepare food as well. Cooking has the effect of making animal protein more readily available for human use and of breaking down starch granules of cereals so that they are more easily digested. The acquisition of fire altered man's eating habits. It made available to him foods that in the raw state were scarcely edible or even toxic. The improved flavor that results from cooking surely must have been appreciated then as now.

Food production also brought changes to man's eating habits and new problems with it. It has been postulated that many people changed from a primarily meat diet, rich in protein, to one comprising largely cereals, which are mostly carbohydrate. If the dependence on cereals were too great, deficiency diseases could have developed, but probably most of the early agriculturists were still getting enough meat, from either wild or domesticated sources, that there was no serious problem until more recent times. Other diseases may have come with the development of agriculture. Storage of food would have brought rats, which can carry diseases to which man is susceptible, and some diseases, such as anthrax and brucellosis, of domesticated animals can be transmitted directly to man. People living in concentrated populations in newly developed urban areas would have been subject to epidemics.

---

*Sometimes, of course, the experiment ended in death when he sampled too much of a poisonous plant. At other times, for example after eating certain kinds of mushrooms, the result may have been unexpected: he would have found himself having hallucinations. Plants causing such effects may then have been put to repeated use, either because man attached religious significance to the effects or simply because he liked them. Some plants may have been used in ways other than for food before a food use was acquired. Hemp, *Cannabis sativa*, is used by man for its fiber, for its edible oil, and, as marijuana, for its euphoric effect. Which of the uses was first acquired is not known.

Food problems are still very much with us today. Some of these, such as hunger, will be the subject of the last chapter. Some of the problems stem from the fact that many foods are highly refined today. It is no longer true that if a man eats what he likes, he eats what he needs, for in many parts of the world today a person may eat an abundance of what he likes from the food available to him and not receive adequate nutrition. Life expectancy has shown a dramatic increase in the last century, and while some of this is due to improved nutrition in spite of the availability of the highly refined foods, more of it is to be credited to advances in medical science.

Many people in the world have been exposed to the rudiments of nutrition in school. They learn, or should learn, the following basic information. The nutrients are carbohydrates (starch and sugar), lipids (fats and fat-like substances), proteins, minerals, vitamins, and, of course, water. Proteins are composed of amino acids. There are some 20 naturally occurring amino acids but only eight, the so-called essential amino acids, must be supplied in our food because the others can be synthesized in the body. (Table 3-1 lists the essential amino acid composition of some foods.) Carbohydrates, lipids, and protein are used to supply energy. Protein, minerals, and water are necessary for building and maintaining the body. All of the nutrients, except carbohydrates, are important in regulating the physiological processes of the body. Primitive man, of course, lived and reproduced without any of this information and many people today, it must be admitted, also get along very well without it, but others suffer as a result of not having—or not heeding—information about nutrition.

Meat is usually man's favorite food and properly so, for meat products provide man a protein comprising all of the essential amino acids. Plants, of course, contain protein but the amounts are usually small and none has a "complete" protein. Moreover, only animals supply vitamin B_{12}, which is necessary in small amounts for proper growth, production of red blood cells, and functioning of the central nervous system. Meat, however, is an expensive food in that animals themselves consume large amounts of food before it is their turn to appear on the dinner table. Thus we could feed seven times as many people directly on grain as are fed when a given amount of grain is converted into meat. This is partly offset, theoretically at least, by the fact that grazing animals are able to utilize food which man cannot. But in the United States today most meat animals are fed grain; 78 percent of the grain produced in the United States is fed to animals. It is only

Table 3-1
Essential amino acid composition (milligrams of amino acid per gram of nitrogen) of certain foods. Egg is considered to have a nearly ideal protein and the other foods are rated in comparison with egg to give a protein score: Note that protein score is not the same as protein content, which is not included in this table.

Food	Iso-leucine	Leucine	Lysine	Methio-nine	Phenylal-anine	Threo-nine	Trypto-phan	Valine	Protein Score
Hen's egg	393	551	436	210	358	320	93	428	
Beef	301	507	556	169	275	287	70	313	69
Cow's milk	295	596	487	157	336	278	88	362	60
Chicken	334	460	497	157	250	248	64	318	64
Fish	299	480	569	179	245	286	70	382	70
Corn	230	783	167	120	305	225	44	303	41
Wheat	204	417	179	94	282	183	68	276	44
Rice	238	514	237	145	322	244	78	344	57
Beans	262	476	450	66	326	248	63	287	34
Soybeans	284	486	399	79	309	241	80	300	47
Potatoes	236	377	299	81	251	235	103	292	34
Manioc	175	247	259	83	156	165	72	204	41
Coconut	244	419	220	120	283	212	68	339	55

SOURCE: Data from FAO Nutritional Studies, No. 24. Rome, 1970.

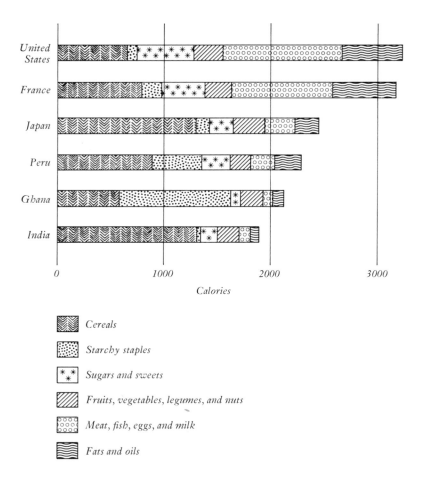

Cereals

Starchy staples

Sugars and sweets

Fruits, vegetables, legumes, and nuts

Meat, fish, eggs, and milk

Fats and oils

Figure 3-1
Sources of food energy in selected countries. Although cereals provide more than 50 percent of man's calories for the world as a whole, their significance varies considerably from country to country. In the more highly developed nations animal products provide a larger proportion of calories for man than do the cereals. Japan is an exception. In tropical African countries the starchy crops such as yam, manioc, and plantain, furnish the greatest source of energy. The net food supply (calories) per person per day also differs greatly between the developed and the developing nations. The daily requirement of an "average" person is estimated at between 2400 and 2500 calories. (Based on data from *FAO Production Yearbook*. Rome. 1969.)

in the wealthy countries that people consume large amounts of meat. In the United States the average per capita consumption of meat is 164 pounds a year, in New Zealand 240 pounds, while in the Far East the average consumption is only a few pounds a year.

There are, of course, many people who, for religious or other reasons, do not eat meat. They are sometimes divided into vegetarians, who eat eggs and milk products, and vegans, who do not eat food of animal origin. The concern over the possible connection of high cholesterol intake with heart disease has strengthened the arguments for a plant diet, since many animal products have high cholesterol levels. Before the advent of plant domestication, however, many people must have subsisted almost entirely on an animal diet, as some Eskimos still do. There are some who still maintain that the best diet for man is one consisting entirely of meat, or nearly so. Man needs certain nutrients, and in the final analysis whether these come from plants or animals is probably largely immaterial.

Many taboos have grown up concerning the eating of animals; the intake of plant foods seems not to have been regulated in this way. People may think there is a rational origin for their rejection of certain foods, but most food taboos actually arose in prehistoric times and their origins will probably always remain obscure. Most taboos are associated with religion but in no consistent way. Some religions permit the eating of sacred animals while others forbid it. It could well be that many taboos did not have their origin in religion but only later became associated with it. Food taboos are still very much with us. In fact, they may at times contribute to protein malnutrition of some people, who will not eat fish, eggs, or other foods because they are "unclean," or thought to be objectionable in some other way.

Much nonsense has been written about food, and, in fact, continues to be written. That much of this nonsense is believed is indicated by the fact that the people of the United States have spent an estimated 500 million dollars a year on food nostrums.

4

Meat: the luxury food

*Let him not eat of either the cow or the ox; for the cow
and the ox doubtless support everything here on earth.
The gods spake, 'Verily, the cow and the ox support
everything here: come, let us bestow on the cow and the ox
whatever vigour belongs to other species!' Accordingly
they bestowed on the cow and the ox whatever vigour
belonged to other species; and therefore the cow and the
ox eat most. Hence, were one to eat of an ox or a cow,
there would be, as it were, an eating of everything, or,
as it were, a going on to the end. . . . Nevertheless
Yâgñavalkya said, 'I, for one, eat it, provided that it is tender.'*
Satapatha-brâhmaṇa III, 1, 2, 21

The domestication of all of our important animal species occurred
quite early. That other animals which might have been domesticated
were not is probably to be partly explained as a geographical and
historical accident. Most of the important domesticated animals came
from the Near East; a few came from southeast Asia. After they were
domesticated, use of them spread around the world. The ancient
Egyptians did keep a large number of other animals, but with the
exception of the cat, none became truly domesticated. In the New
World a few animals were domesticated, but with the exception of
the turkey, none were to become widely used outside of their home-

land. Columbus brought cattle and sheep with him on his second voyage and the Old World animals soon became widespread in the Americas. Man has been content to try to adapt the same animals to new habitats rather than to exploit new species, although recently there have been a few attempts to domesticate other animals, such as the musk ox. The eland, an antelope, is now said to be truly domesticated in Africa as the result of such an attempt.

A domesticated animal is one that breeds under man's control. (If we accept this definition, it then follows that man is not a fully domesticated species, for he has not yet succeeded in controlling his own breeding.) This definition might be quibbled over, but it should serve. An animal may be tamed as a wild plant may be cultivated but that does not make either of them "domesticated." Some animals whose feeding and protection have been assumed by man have become so completely domesticated they are no longer able to survive without man's care. Other domesticated species may occasionally become feral, or wild, as the horse did in the western United States.

Present evidence indicates that plant and animal domestication began at approximately the same time in the Near East although it has frequently been assumed that domestication of plants preceded that of animals. The earliest date known for domesticated sheep antedates those known at present for any cultivated plant. It may have been that some hunters and gatherers were domesticating animals while others were concentrating their attention on plants. Some have argued, however, that even if this were true, before domestication of animals could have proceeded very far, plant cultivation and village life would have to have been established. Thus the great division of farming and pastoral people—reflected in the Old Testament story of Cain and Abel—may have been not an early development but a later one. Future archaeological work, of course, may shed more light on whether plants or animals were the first to be domesticated.

Only about 50 or so animals have been truly domesticated, if the honey bee, the silk moth, and a few aquatic animals, such as carp and trout are included. Only a dozen or so of the domesticated animals are of great importance and have a wide distribution. If we omit dogs and cats, for which there are no reliable estimates, we find that chickens are the most numerous, with an estimated three billion for the world. Sheep are the next most numerous species, having about one billion representatives, and cattle are third. Pigs, goats, and water buffalo are also numerous.

Today in many parts of the world animals are still kept in ways not very different from those of the prehistoric period, but there have been radical changes in the United States and some of the other developed nations, most of them dating back only a quarter of a century or so. The small barnyard with its variety of animals is disappearing and is being replaced by large specialized farms. There are now automatic feeding devices that make use of the latest nutritional research. It is probably correct to say that we have a better understanding of the nutritional requirements of some of the domesticated animals than we do for man. Protein concentrates and feeding supplements are extensively used. Modern sanitation practices, vaccination, and treatment of diseases with antibiotics and other drugs have contributed to the production of much healthier animals. The animals themselves have changed as a result of work done at modern breeding farms and under planned breeding systems based on extensive records and supervised by trained geneticists. Animal breeding today is based on performance characteristics such as amount and quality of milk, eggs, or meat produced, rather than on appearance as judged at state fairs or animal shows. The new developments, coupled with highly organized processing and marketing systems, have drastically changed the livestock industry in the space of a few years.

Just as they do today, animals served man in early times in many ways other than as food. They provided leather and wool for clothing, bones for tools, dung for fertilizer and fuel, a means of traction and transportation, as well as serving for amusement and religious offerings. Today they are also used in the manufacture of pharmaceuticals, fuel, fertilizers, greases, oil, gelatin, glue and other industrial products. Catgut, usually made from sheep intestines, is used for violin and tennis strings and sutures. Chiefly because of a large number of synthetic products, however, we are far less dependent on animals for raw materials than was primitive man. They also serve us today as subjects in experiments for medical and scientific research.

One group of animals (the order Artiodactyla), the even-toed ungulates or hoofed animals, has furnished man with 15 of the 22 most important domesticated animals. Of these the ruminants, or cud-chewing animals, include three of the most important as well as earliest domesticates, cattle, sheep, and goats. These animals, because of the microflora—bacteria and protozoans—present in their stomach, are able to digest food that man cannot and thus to convert second-rate protein into first-rate protein for man. In a sense they were preadapted to domestication in that they were not in competition with man, being able to exist on a diet that was utterly worthless to him. Also the fact that they were social rather than solitary animals may have made their domestication more easily accomplished. Man's most important food animals will be treated in greater detail in the following paragraphs. The first to be discussed, seldom used for food today, may be man's oldest domesticate.

Dogs *(Canis familiaris)**

Although the dog has generally been regarded as man's first domesticated animal, this has not been proved by archaeology. The oldest known dog remains have been found in North America and date back to about 8400 BC; the oldest remains found to date in the Old World—some are from western Europe and some are from the Near East—are about a thousand years younger. Nevertheless, it has generally been assumed that the dog was domesticated in the Old World and came to America with man in late migrations across the Bering Strait. It now seems fairly clearly established that the wolf was the ancestor of the dog. Wolves had a wide distribution in both the Old and New World and still have a fairly extensive distribution. The place or places of origin of the dog is far from certain—some people have suggested the Near East, others India. It is, of course, not improbable that more than one domestication could have occurred. Wolves are easily tamed when young and perhaps various groups of men brought wolves to their camps at different places and times and subsequently domesticated them. Dogs have a curly or sickle-shaped tail in contrast to the drooping tail of their wild relatives. Selection for this character would perhaps have helped man in distinguishing his own animals from wild ones. The mutation or mutations which gave rise to this character need not have happened only once.

How important dogs were to hunters in early times is the subject of some dispute; some persons think that they played no great role.

---

*The scientific name of a plant or animal is composed of two words, the genus name (*Canis*, in our example) and a modifying epithet. Thus *Canis familiaris* is a species, or specific kind of organism, that is commonly called the dog. A genus is composed of one or more species that have several characters in common. Species may be subdivided into varieties or subspecies. Genera in turn are grouped in families, families into orders, and so on.

Figure 4-1
Clay dog, Tolima, Mexico. Dogs for eating were
fattened on maize. (Original in National Museum
of Anthropology, Mexico.)

Nor is there agreement about whether dogs were an important aid to
man in helping to round up and control the herd animals that were
his next domesticates. In times when food for man was in short supply, the dog would have been a competitor. The dog is a scavenger,
however, and in time this trait became appreciated; it is still the dog
that helps to keep the village clean in many areas of the world. In
parts of both the Near East and South America today dogs feed
largely on human feces. The dog became prized as a source of food
in many cultures, and in Mexico a special breed grown for eating was
castrated for fattening. In China the dog was used for food until
quite recently. According to report, dogs are now being eliminated
in that country because they compete with man for food. Although
dogs may not be as common in China as in the past and may have
disappeared as pets in many regions, as recently as 1963 they were
observed to be common in Outer Mongolia where they were being
used to guard the flocks from wolves.

The dog became the most widely distributed domesticated animal
because he was appreciated and respected as an aid in hunting, in
keeping the flocks, and as a scavenger, as well as being used occasionally as a draft animal or as a source of fur or food. Although in some
parts of the world the dog still earns his keep in these various ways,
his role is coming more and more to be that of man's favorite pet. As

such, his contribution to pollution in larger cities has been news-worthy in recent years.

Sheep *(Ovis aries)*

The ruminants of the family Bovideae include sheep, goats, cattle, and the water buffalo—animals valued highly not only for meat, but for milk and for their skins or wool. Sheep are the first of these animals to appear as domesticates in the archaeological record, being represented by remains dated back to 9000 BC.

At one time wild sheep, known as urials and mouflons, were widespread across most of Asia, and exactly which race or races provided stock for the origin of the domestic forms remains somewhat speculative. The bighorn sheep *(Ovis canadensis)* of Asia and North America was never domesticated. Today most of the wild sheep of the Old World are mountain animals, although they probably existed in lower regions in earlier times. Most of the bones uncovered come from the Near East, and it is not always clear whether the bones found in the earliest archaeological deposits are from domesticated or wild animals. With domestication came changes in the ear, the face, the horns (which in some races disappeared entirely), the color, the wool, and the tail—mostly traits of little help to the archaeologist for distinguishing the bones of wild and domestic forms. A large tail with an abundance of fat was much appreciated by early people, and apparently there was deliberate selection for this character. Some sheep developed tails so heavy that carts were constructed to help the sheep carry them around. Fat-rumped sheep were also selected in early times because animal oils were highly valued for use in lamps. Wool may have been made into felt cloth long before it was spun and woven; perhaps the spinning and weaving of wool was first done in places where such plant fibers as flax were not available. Sheep are unique among domestic animals in their adaptation to environments that are extremely unfavorable for other animal species. They probably became more widespread than goats in early times because they could survive better in hot climates. They have a panting mechanism that allows them to tolerate hot climates, and, surprisingly, the wool has also been suggested to function as a cooling device in sunny desert areas although most authorities consider it only an adaptation for cold.

Figure 4-2
Sheep in Bolivian highlands. (Courtesy of FAO.)

Figure 4-3
Fat-tailed sheep with cart for carrying tail from Rudolf
the Elder's *New History of Ethiopia*, AD 1682. (From
Frederick E. Zeuner, *A History of Domesticated
Animals*. Hutchinson, London, 1963. Used by permission.)

Today, as is true for nearly all of our domesticated animals, there are numerous breeds of sheep, some specialized as producers of wool, others for meat, and still others for milk. Most breeds of sheep in the United States, however, are used both for wool and meat. Cattlemen are often opposed to sheep production because the grazing habits of sheep often destroy the rangeland and their fine hooves tend to ruin the watering places.

Goats (*Capra hircus*)

Although dated remains of the first known domesticated sheep are 2000 years older than those of goats, this does not necessarily indicate that the goat was domesticated later. In many deposits bones are present that could have belonged to either goats or sheep; positive identification is not always a simple matter. In many places where remains of both animals are found in the same deposits, the goat remains appear earlier than those of sheep. Wild goats or bezoars, the ancestral form, now quite rare, once extended across southern Asia from India to Crete. The original domestication could have been in the Near East, both Persia and Palestine having been suggested. The goat, in contrast to the other domesticated animals, is a browser, even at times climbing trees to get at leaves, and can survive in areas where the food supply is inadequate for other animals. It also is able to graze; it may eat grass so close to the roots as to promote erosion in seasonally dry areas. There is some controversy, however, about how detrimental the goat is to the land. Although the goat has been generally condemned for his destructive grazing habits, he has also had his defenders who point out that goats are often put to pasture where cattle have already fed, and thus at times may be blamed for erosion actually started by the cattle. It has, moreover, been pointed out that shrubs may invade and ruin grasslands overgrazed by cattle but that goats protect such grasslands by eating shrubs.

Domesticated goats spread rapidly but were nearly always less appreciated than sheep, which are far superior for wool production, and cattle, superior for milk production; both sheep and cattle were generally preferred over goats for meat. Goat production is not very significant to man today, except in certain steppe and mountainous areas and in parts of Africa and Asia where goats are still more important than sheep. At present nearly half of the world's goats are

48

Figure 4-4
Goat browsing on tree. (Courtesy of FAO.)

found in Africa. One breed, the Angora goat, is still important for its wool, which is called mohair. Goat's milk, which in some ways is more similar to human milk than is the cow's, still has some use. The fat globules are small in size and more easily digested than those of cow's milk. It is rather difficult, however, to imagine people who had either sheep or cattle domesticating goats, which might perhaps argue for goats' being the first ruminants to be domesticated or for their being domesticated in areas somewhat removed from centers of sheep and cattle domestication.

Cattle *(Bos taurus)*

What most people would consider the aristocrats of the domesticated animals, cattle, so far as we know today, were domesticated later than sheep and goats. Cattle first appear in the archaeological record shortly after the first appearance of goats—but they do not become common in the archaeological record until much later. Whether success with other animals inspired attempts at domestication of cattle or whether they were domesticated by people unacquainted with other domesticated animals is not known. The aurochs, or wild cattle, were worshipped long before their domestication, and a religious motive for their domestication has been postulated as was mentioned in a previous chapter. The animals, of course, had been hunted in earlier times, as some magnificent cave drawings in Europe attest. The aurochs, the last of which were killed in Poland around AD 1630, were magnificent animals. Cattle, probably rather similar to some of the original wild animals, have been developed in modern times in Germany by interbreeding modern types with certain presumably primitive characters. The animals ultimately produced were large and strong, temperamental and ferocious, and quite agile, unlike the thickset beasts most common today. The domestication of the aurochs was certainly not a simple matter. One of the changes frequently accompanying domestication in animals is a decrease in size from that of the progenitor. In cattle this may well have resulted from an intentional selection by man of smaller beasts since they might be more easily managed. However, it could also simply be due to an environmental factor in that the animals may not have fed as well under man's control as they had in their natural environment. The aurochs were widely distributed throughout the temperate parts of Europe, Asia, and North Africa. The earliest known remains of domesticated cattle, dated at 6300 BC, come from Greece, and they are known from Anatolia at 5800 BC. It seems probable that there was a separate domestication of humped cattle in India, probably from a native species known to have been there since the Pleistocene. Humped cattle were present in Mesopotamia at about 4500 BC, which, if a domestication in India is accepted, implies early contacts between this area and eastern Asia. Crosses between the two types of cattle occurred then as they do now. Various breeds developed quite early. As man introduced cattle into new areas there was probably mating of the cows with wild bulls, either intentionally or unintentionally. Primitive man is thought to have staked out female animals, as is still practiced with reindeer, to entice male animals, which are then captured or killed for food. The introduction of genes from mating of the wild bulls with the domesticated stock probably con-

Figure 4-5
The aurochs, based on a drawing of the last surviving specimen. (From
Frederick E. Zeuner, *A History of Domesticated Animals*. Hutchinson,
London, 1963. Used by permission.)

tributed to the early development of considerable diversity in cattle
and was the foundation of new breeds.

When man first found that castration* could have a profound
effect on the bull, rendering it docile and manageable, is not known.
A religious origin, as a sacrifice of the male element, among people
who worshipped cattle, has been suggested, but perhaps it is more
likely that it was a purely practical matter. Man would have found
that keeping more than one bull in a herd created difficulties that

---

*Man also quite early practiced castration on other animals including himself.
The most widely practiced method, now as then, is to open the scrotum with a
knife and remove the testicles, but some people castrated their animals by
pounding the testicles between stones. In addition to changing the metabolism
and behavior of the male animals, castration obviously was a method of birth
control. There were probably good reasons at times to restrict the breeding of
animals, such as the fact that young born at certain times of the year in adverse
environments might have little chance of survival. Other methods of birth control included fitting leather aprons to the animals or binding their prepuces with
string. Both of these methods are still used with rams in Africa.

could be solved by castrating all but one of them. The ox, which was one of man's first beasts of burden, is still very important in parts of the Old World and Latin America. It made possible the plowing of fields, the plow perhaps coming into use around 3000 BC, and thus had a profound effect on the development of agriculture by considerably extending the area that could be tilled.

Man must have used milk for food soon after the herd animals were domesticated. Although milk is still generally regarded as one of man's best foods, there are many adult people who cannot tolerate it, as was learned when the United States sent powdered milk as part of relief shipments to various countries. Lactose, or milk sugar, cannot be directly digested by man but must be broken down into simpler sugars by an enzyme called lactase. Virtually all babies have lactase, which enables them to digest their mothers' milk. In the past it was thought that humans lose the ability to digest lactose if they are not continually fed milk after weaning, but it has recently been pointed out that there may be a genetic difference among people for the ability to digest lactose. It is largely adults of the nondairying cultures of Africa and southeast Asia and of American Indian groups, who are unable to digest lactose. These people, however, can eat cheese, which has a very low lactose content.

In parts of Africa in which they became a symbol of wealth, cattle were used for currency, and the bride price is still frequently paid with cattle. In some areas of Africa cattle are seldom or never eaten, although the blood and milk may be used for human food. The Masai tribe obtains the blood by shooting an arrow at close range into the vein on the neck of an animal, and then collecting the blood in a gourd. The Abahima of Uganda have selected their cattle for large horns, which may reach weights of 150 pounds, and a hump so large that it droops over to one side. The appearance of these animals is pleasing to the Abahima but the large horns and humps represent no economic benefit and, in fact, may be detrimental to the well-being of the cattle.

The role of cattle in India has received considerable publicity. Cattle are sacred among the Hindus, and it has been said that they would rather die of starvation than kill their animals. Their failure to use cattle as a source of food has been much criticized by outsiders. The origin of the taboos on eating cattle in India is obscure. That the animal was sacred may or may not have much to do with it, for some

Figure 4-6
Zebu. These cattle, which have been found to be superior to other cattle in many tropical and subtropical areas, have been widely used for breeding purposes. (Courtesy of FAO.)

peoples are known to eat the animals they hold sacred. It has been suggested that the animals were so valuable as a source of traction and in providing milk that there was an early prohibition on killing them. Later, of course, the taboo may have been reinforced with the entry of cattle-eating foreigners, first the Moslems and later the British. Estimates on how many people will eat meat in India vary greatly. One recent account has stated that 70 percent of the Indians are non-vegetarians, another states that 90 percent are vegetarians! There is a suspicion that few of the cattle that die do not end up in a pot somewhere. India is overpopulated with cattle—as are parts of Africa, but for other reasons. One-fourth of the world's cattle are found in India. The animals do make many contributions. The cows provide milk, the bullocks are the principal source of traction, the dung is used for cooking fuel as well as for a construction plaster, but the facts remain that it takes fifteen cows to produce as much milk as one does in the United States and that the Indians cannot get efficient labor from ill-fed bullocks. It has been pointed out that the Indians would be far better off with fewer, more productive animals even if they were not used for meat. The subject of cattle is still a highly emotional issue in India.

The Indian, or zebu, cattle, sometimes known as Brahman cattle in the United States, in addition to having humps, differ from other cattle in having long heads with drooping ears, loose skin and longer legs, and in being well adapted to hot climatic zones that have a pronounced dry season. They have been extensively used in this century for breeding purposes in the southern United States and in Latin America to provide more productive cattle for these areas.

The value attached to cattle as food for man has led to their introduction into many parts of the world where they are not well adapted, as for example, the semiarid scrub region of the western United States. Cattle cannot be grown in some parts of Africa because of parasites, chiefly tsetse flies. Various diseases still plague cattle in other parts of the world, of which hoof-and-mouth disease is one of the most serious. An outbreak of this occurred in England in 1967, and more than 400,000 cattle had to be destroyed and burned, representing a loss of $250,000,000.

Cattle in the United States today are mainly bred for milk or beef, with only 15 percent being dual purpose. Improved breeding and feeding practices and modern transportation have made cattle far more efficient meat producers than in the past, but beef still ranks as man's most expensive meat in terms of the cost of feeding cattle. There have also been great advances in dairy science in recent years, milking machines, for example, are now widely used; and artificial insemination and the freezing of semen are extensively employed in breeding. High butterfat yield, once considered very important, is no longer emphasized because of the present concern over calorie and cholesterol intake as well as the fact that a large number of plant oils are available for butter substitutes. In addition to cattle, the genus *Bos* has supplied man with three other domesticates in Asia, the mithan *(Bos frontalis)*, Bali cattle *(Bos javanicus)*, and the yak *(Bos grunniens)*.

Figure 4-7
A dairy herd of water buffalos near Calcutta. It is necessary for the water buffalo to spend considerable time in water. (Courtesy of FAO.)

Water Buffalo *(Bubalus bubalis)*

Another member of the bovid family—the water, or Indian, buffalo—is still extremely important in many parts of the world. The other buffalos and bisons of Europe, North America, and Africa were never domesticated. We know little more about the domestication of the water buffalo other than that it occurred in India presumably sometime before 2500 BC. The water buffalo thrives in tropical low-

Figure 4-8
Water buffalo plowing a paddy field in Burma. (Courtesy of FAO.)

lands, likes to wallow in water, and lives largely on aquatic or semi-aquatic grasses and other vegetation. Thus it is adapted to areas and foods that will not support cattle, which may explain why it is still so important today. Its use is similar to that of cattle, and it gives more milk than most breeds of cattle, with an 8 percent butterfat content, twice that of cow's milk. Today it produces about half of India's milk supply. Butter made from the milk is greenish-white in color, is more solid than cow's butter, and turns rancid less readily. Its meat has a flavor very similar to that of beef but it has a distinct bluish tinge and fat that is white. Thus far there has been little scientific effort to improve the animal.

The water buffalo spread throughout most of southeastern Asia where it became important in preparing fields for growing rice. It is, in fact, the only animal well adapted to work in muddy fields. The late arrival of the animal in Africa and Europe is rather puzzling. It didn't reach Italy until about AD 700, where today its milk is used in making mozzarella cheese. It is also raised in Romania, Bulgaria, and Egypt. It was brought to Brazil in 1903, where it has become important in the lower Amazon valley. Efforts to introduce it in many parts of Africa have failed. The animals have escaped in northern Australia and become feral as also has happened in Brazil and in their homeland.

Reports of hybrids between water buffalo and cattle are sometimes made but have not been scientifically verified. On the other hand, hybrids between different species of cattle *(Bos)* are not uncommon in Asia, and hybrids, known as cattalo, have been produced between domestic cattle and the American bison or "buffalo."

Horses *(Equus caballus)*

Most readers of this book will have never eaten horse meat, or at least, not knowingly so. Occasional "scandals" have stirred in the past when it has been found that horse meat was substituted for beef. There is nothing wrong with horse meat: wild horses were common game for our Paleolithic ancestors, and horse meat is still eaten by many people today in inner Asia, which is the area where the horse was domesticated. The milk of the horse is employed for making kumiss and other fermented beverages. Why then do so many people fail to eat horse flesh? As with all food taboos, the origin of the avoidance is obscure, and there probably is no logical reason. Various suggestions have been put forward. The horse, more than any other animal except dogs and cats, was man's friend and close companion and man doesn't usually eat his close friends. Perhaps the animal was so useful to man in other ways, in agricultural work and transportation, or in warfare that its eating was discouraged. Possibly Christianity played a role in the rejection of horse meat in Europe, since the eating of it was associated with pagans. There were attempts to popularize it in Europe in the last century, but they largely failed except in France. In 1969 quite an outcry was raised in England when the people learned that retired horses from the Queen's pound were being sold in France for use as food. As a result the Government changed its policy, promising to put retired horses out to pasture. It seems unlikely that more widespread use of horse meat would do

Figure 4-9
Donkeys used to haul grain. From an Egyptian tomb, c. 2400 BC.
(From Charles Singer et al., eds., *History of Technology*, Vol. I.
Oxford University Press, London, 1954.)

much to ease hunger in the world since horses, like cattle, are expensive eaters.

There is no good archaeological evidence providing clues about the time and place of the domestication of the horse. It is thought to have occurred in the steppe area of what is now the southern Russian Turkestan region. The tarpan, which became extinct in the last century, was probably its chief wild progenitor. The domestication of the horse, which was to benefit man in many ways, unfortunately had other consequences. Domestication had probably occurred by 4000 BC; and once they were provided with riding horses, the nomadic peoples became a great scourge to the sedentary farmers. From 2000 BC on mounted warriors and horse-drawn chariots "swept across the western world," as Zeuner expresses it. The horse made it possible for the Huns to build their great empire and it promoted the many martial successes of the Arabs. The horse was responsible for the successful invasion of Genghis Khan and it helped the Spanish to conquer the Americas with great rapidity. Until this century the horse continued to occupy a major role in wars. Two other late domesticates, the camel and the elephant, were also used in warfare, but neither was as important as the horse.

Two other species of *Equus* were also domesticated, the onager or half-ass, and the ass or donkey, but neither of them was to become quite as closely associated with man as was the horse. The mule, which results from the mating of a mare and an ass,* was to become the first documented interspecific hybrid on record. The mule has been described as "an animal without pride of ancestry nor hope of descendants," which isn't quite true, for there are a few records of mules producing offspring. Throughout history the mule has been cursed for its disposition but praised for its surefootedness and has made its contributions to agriculture.

In 1950 it was estimated that 86 percent of the draft power for the world's agriculture, and 25 percent for the United States, was provided by animals. The estimate seems too high for the United States, and, of course, the use of animals for draft power has drastically declined in the United States since that time, although for the world as a whole it is decreasing much more slowly. In 1918 there were 27 million horses in the United States. This figure declined to 3 million in 1960 at which time the census of horses was discontinued. The total number of horses, however, is now thought to be somewhat higher than in 1960, for although their use in agriculture declines every year, their use in recreation and sport is on the increase. The horses don't always come cheap. In recent years there have been reports of race horses that have sold for more than a million dollars.

Pigs *(Sus scrofa)*

Pigs, like dogs, are scavengers and hence were in more direct competition with man for food than were the grazing animals. Also, unlike many of our other domesticated animals, they do not have an important dual role, the pig being kept almost solely as a meat producer. The prolific pig was, and still is, a wonderful supplier of meat, the most productive of food per unit area of the larger domesticated animals. Today it ranks next to beef among the preferred meats in the United States, and in some parts of the world it is the principal meat source. It has been said that in the modern meat industry every part of the pig is used except the squeal. Its hide, of course, has long been appreciated and "pig-skin" is synonymous with football in the United States. The scavenging habit of the pig may have been important among primitive people in preparing the land for farming, and

---

*The reciprocal hybrid, from the mating of ass with stallion, is called a hinny.

Figure 4-10
Sheep used to tread in seed. The sower (right) offers grain to the animals to keep them following him. From an Egyptian tomb, c. 2400 BC. (From Charles Singer et al., eds., *History of Technology*, Vol. I. Oxford University Press, London, 1954.)

in fact, pigs probably helped open the forests of Europe for crop planting. The ancient Egyptian employed pigs or sheep to tread seed in the ground for planting. At times it has been used for traction. One of the most interesting specialized uses is that pigs, like dogs, have been trained to hunt for truffles in France, and it has even been reported that some have been used for retrieving game.

The origins of the domesticated pig are perhaps even less clear than those of most of the animals previously discussed. Wild pigs are native from Europe to eastern Asia and it appears that there have been at least two separate domestications, if not several, the European pig having come from the wild *Sus scrofa* and the Chinese pig from *Sus vittatus*. However, there may be little justification for regarding the ancestral forms as separate species, and perhaps they are better considered geographical races or subspecies of one wide-ranging species. Again it appears that the earliest domestication may have been in the Near East, for pig remains dated back to about 7000 BC are known from Turkey. No date or locality can be postulated for the origin of the Chinese pig. In all probability, however, its domestication was later than that of the pig in the Near East. There may have been many local domestications of pigs in different parts of Europe, for the pig probably was not a particularly difficult animal to bring into domestication.

The domestication of the pig thus is somewhat later than that of the sheep, and this may be due to the fact that pigs couldn't have been domesticated until there were well-established villages since the pig is primarily a household animal, not a herd animal. The fact that it could not readily be herded may have led to the development of taboos against it among nomadic people, such as the Jews. To them it could have become a despised animal since it was available only to settled people, who were at the same time people with alien gods. This explanation for the origin of the taboo is not proved, of course, but it seems a more reasonable one than others that have been proposed. It seems very unlikely that primitive people could have known that pork could carry trichinosis, or would have rejected the animal because it was a scavenger. Mohammed had considerable influence on the rejection of pork among his followers, perhaps to make a distinction between his people and the Christians, who were pork eaters. The pig was the most important sacrificial animal in ancient Rome, and it is still important in religious observations in parts of Southeast Asia and the Pacific region where the tusks often have special ceremonial significance. Among some tribes of New Guinea pigs are eaten only on ceremonial occasions.

Just as was the case with cattle in Africa and Asia, religious beliefs may impede the improvement of pigs in some parts of the world. In an attempt to improve the food supply of people in an area of Upper Burma, agricultural agents hybridized the local black pig with a higher-yielding spotted strain. Because the hybrids were spotted, the experiment was a failure, for the local people believed that spotted pigs were unfit for human consumption.

The modern thickset pig most widely raised in North America is derived from crosses of European and Chinese breeds first made in England more than 150 years ago. The smelly pig sty is disappearing today, as large modern farms take over the pork and bacon production in the United States. Corn was found to be an ideal fattener for swine, and therefore it is no surprise that the corn belt of the north central United States is the chief production area of swine. Castration, a very simple operation with pigs, is widely practiced for purposes of fattening.* Boars intended for meat are castrated a few months before slaughter in order to eliminate an undesirable odor from their meat. The breeding of pigs in recent years has produced

---

*The Tsembaga of New Guinea castrate all of their male pigs, which means that they have to depend upon feral males to perpetuate the race.

many changes to meet consumer requirements. People want less fat than formerly, and lard is no longer in great demand.

Chickens *(Gallus gallus)*

Of the domestic fowl, chickens are more important than all of the others combined, and they are one of the few important domesticated animals to have come from the Far East. The chicken is derived from the jungle fowl of India, and it has been postulated that the fowl was originally domesticated to increase its availability for use in divination rather than for food. Attempting to foretell the future by examining the entrails or the perforations of the thigh bone is still practiced in some parts of Southeast Asia. This is also the area where cock fighting apparently originated. Cock fighting was enjoyed in ancient Greece, and is popular today in various parts of the world. How early the chicken was domesticated is not known, but it apparently didn't reach Persia and Egypt before 2000 BC, from where it spread to Europe a thousand or so years later. The cock was early valued, and perhaps also reviled, as a time clock, and its crowing was thought to frighten away the evil spirits of the night. In some areas it became an erotic and fertility symbol, the former because of the cock's elaborate courtship behavior, the later because of the hen's abundant egg laying. At the same time some people in Africa avoided it because they believed that eating of it or the eggs would destroy their sexual functioning and fertility.

The chicken continues to serve as a dual-purpose animal in many parts of the world; in many places the birds are scrawny, largely scavenging for an existence, and producing few eggs, which are rarely eaten by the owners but taken to the market for sale. More and more, the production of eggs and "broilers" are separate operations in the industrialized nations, where the raising of chickens depends on procedures that offer a striking contrast to the old barnyard methods. In fact, the broiler farm may be compared to a modern factory with its assembly lines. Mammoth incubators, automatic feeding, watering, ventilating and cleaning, and defeathering machinery are utilized in the production of fowl for the supermarket or chicken-house chain restaurant. Breeding of poultry is more highly developed than that of any other group of animals, and hybrid chickens have come to be of great importance in recent years.

A

Figure 4-11
A. Chickens, along with pigeons, turkeys and goats, being fed in a farm yard in Togo. (Courtesy of FAO.)
B. Collecting eggs in a modern poultry farm in Japan. (Courtesy of FAO.)

B

Other Old World Domesticates

There are, of course, several other animals—camels, elephants, reindeer, for example—that are used in the Old World. None of them, however, figured as prominently in early agriculture as did those already treated. Nor are these other animals as important as food sources today as some of those previously discussed. This is not to deny that some of the other domesticated animals—the rabbit, for example, which can be raised very economically—cannot make significant contributions in the future by supplying much needed protein in various parts of the world.

New World Domesticates

In the Americas very few species of animals were domesticated, and none of these, with the exception of the turkey *(Meleagris gallopavo)*, have ever become significant outside of the New World. Wild turkeys, now rare, were once fairly widely distributed in North America and Mexico, and the turkey was a well domesticated animal in Mexico at the time of the arrival of the Spanish. Whether the bird had reached South America in pre-Conquest times is not entirely clear. Turkeys or "pavos" are mentioned in the early Spanish accounts of parts of South America but these references may actually have been to other large birds. Our name turkey, of course, is in a sense an error in that the American bird was confused with the "turkeycock" or peafowl in England and the name became transferred to it.

In South America four animals were domesticated in prehistoric times. The Muscovy duck *(Carina moschata)*—why it is called Muscovy has never been entirely satisfactorily explained—is the other fowl to have come from the Americas. The guinea pig, cavy, or "cui" *(Cavia porcellus)*, a rodent, now used as a laboratory experimental animal, was domesticated in the Andes and is still an important food animal among the Indians there, living in their houses with them and often appearing for sale in the markets, either alive or roasted. The name "guinea" is thought to stem from their arrival in Europe from Guinea, the animals having gone first from South America to West Africa. More important than either of these species, however, are the llama *(Lama glama)* and the alpaca *(Lama pacos)*, both came-

Figure 4-12
Roasted guinea pigs for sale in a market in Quito, Ecuador.

loid animals probably derived from the wild guanaco. A relative of these, the vicuña, much prized for its wool, was never domesticated.

Llama remains have been reported at a site of early agriculture in Peru, dated roughly at between 2500–1250 BC. The original domestication may have been in the Lake Titicaca region, an area of early high cultures in South America. The llama played many important roles among the Andean Indians. The animals were used for sacrifice—pure black or white animals being preferred for this purpose—their lungs and entrails were examined for omens, and potatoes were

Figure 4-13
Llamas in Peru. (Courtesy of FAO.)

ritually treated with their blood before planting. A religious motive for their domestication has, of course, been suggested. The animal was used as a beast of burden, for food, as a source of wool and hide, as a medicine, as a source of dung for fuel, but was neither ridden nor milked. As beasts of burden llamas carry only a relatively small load, a little more than 100 pounds, and they travel 10 to 18 miles a day. A great advantage was that they could live off the sparse fodder provided by the Andean highlands. The alpaca, which is slightly smaller than the llama, is used for similar purposes except not as a beast of burden. Its wool, however, is far superior.

After the arrival of the Spanish, the Old World domesticated animals spread throughout the Americas. In Peru it is sometimes said that it took three other animals to replace the llama—cattle for food, the donkey as a beast of burden, and sheep for wool—and that all of them required more food than the llama. Both the llama and alpaca have persisted in the high Andes where they are so well adapted, and they will probably continue to do so for some time to come. Although their range is considerably more restricted than it was at the height of the Inca empire, they are still a common sight in parts of highland Peru and Bolivia where they are still used much as in the past and are gaily decorated for ceremonial occasions.

5

Grasses: the staff of life

All flesh is grass. Isaiah 40:6.

Of all the various plant groups—algae, fungi, mosses, ferns, gymnosperms, and flowering plants—the last named has furnished us with nearly all of the species we use for food and clothing and in countless other ways. Since the flowering plants, or angiosperms, comprise nearly two-thirds of all species of plants and are the dominant vegetation on the earth's land surface, their great use perhaps should not come as a surprise; but their significance derives not so much from their number as from the fact that they are the only group that produces fruits and seeds. Of the 200,000 or more species of flowering plants known, only 3000 or so have been used to any extent by man for food, and of these about 200 have become more or less domesticated, of which only 12 or 13 are of major importance. Of these dozen or so species, the grasses* contribute four—wheat, rice, maize,†

---

*Quite a number of plants are sometimes called grass, including marijuana, that do not belong to the botanical family that carries this name. The true grasses are distinguished from other plants by a combination of floral and vegetative features. Their small flowers, which lack petals, are enclosed in specialized scales or bracts. The flowers are grouped in small clusters or spikelets, and the spikelets, in turn, are arranged in clusters, such as a head of wheat or a corn tassel.

†Maize is a name of the plant that is commonly called corn in the United States. In other English-speaking countries, "corn" generally means the most common cereal or sometimes any cereal. In the Bible "corn" usually refers to wheat as it does in much of the United Kingdom today, except in Scotland where it refers to oats.

and sugar cane. Of the 300 or so families of flowering plants none is of greater importance to man than is the grass family, known scientifically as the Gramineae.

From the time of the earliest seed collectors to the present the grasses have supplied man with one of his principal foods. Their fruits, or grains, often called seeds, each develop from the single ovary of the flower and each contains a single true seed. Inside the seed coat is a rich layer of stored food, mostly starch, known as the endosperm. This concentrated reserve food, which is designed to provide energy for the germinating embryo, is also a rich source of carbohydrates for man and other animals. The embryo, or germ, of the seed contains protein and oil. Some vitamins and minerals are also present in the grain. Thus the grasses come close to being the ideal source of plant food for man but alone they can not sustain him, for their protein does not contain all the amino acids essential for man's well being. The cereals are also deficient in calcium, and except for the yellow form of maize, in vitamin A, and the dried seeds do not contain vitamin C. The grains give high yields, are fairly easy to collect, and may be stored for long periods of time without spoiling. Little wonder that the cereals—so named from Ceres, the Roman goddess of crops—have become the chief crop of most men throughout the world. More than 70 percent of farm land is planted to cereals, which provide man with more than 50 percent of his calories.

Wheat and barley formed the basis for the development of early civilization in the Near East, rice was the food that allowed the development of high civilization in the Far East, and maize was responsible for the development of the high cultures in the Americas. These four grasses, along with oats, and rye, are the cereals best known to people of the Western world. In some parts of the world, particularly in Africa and Asia, other grasses, better adapted to climates too severe for the major cereals, are grown. The most important of these is grain sorghum. A number of different grasses, sometimes lumped under the name "millets," are also fairly extensively grown as cereals as is Job's-tears, perhaps best known in the United States in the form of beads used for necklaces. All of the grasses thus far mentioned, with the exception of maize, are native to the Old World, and it is sometimes stated that maize was the only true cereal ever domesticated in the New World. This is not quite true, for in prehistoric times Indians in southern South America cultivated a brome grass for its grain and in parts of Mexico a species of panic grass was grown but neither of

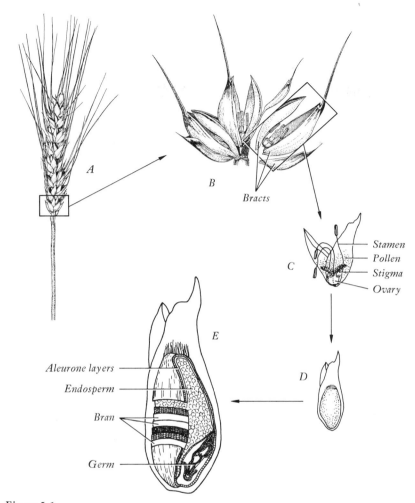

Figure 5-1
Flowers and grain of wheat. *A.* Head or spike composed of many spikelets.
B. Spikelet with three flowers. *C.* Single flower showing parts. *D.* Maturing
ovary or young grain. *E.* Mature fruit or grain, surrounded by bracts or chaff,
dissected to show various parts. The outer layers of the grain, the bran, contain
some carbohydrates, B vitamins, and minerals. The outer part of the endosperm,
the aleurone layer, contains protein and phosphorus. The endosperm, or food
layer of the seed, is composed mostly of carbohydrates; this is the only part used
in making highly refined flour. The germ or embryo of the seed is rich in fats,
protein, and vitamins, and contains some minerals and carbohydrates. (*A-D*
from "Hybrid Wheat" by Byrd C. Curtis and David R. Johnston. Copyright ©
1969 by Scientific American, Inc. All rights reserved. *E* from Wheat Flour
Institute.)

Figure 5-2
Bamboo plants. (Courtesy of FAO.)

them was to achieve any importance. Quite apart from the use of grains for food, grasses serve mankind in many other ways. Sugar cane is the source of 60 percent or more of the sugar produced in the world.

With the coming of civilization man needed something to bring relief from the worries and cares that accompanied it. For this he often turned to the same plants that had played a fundamental part in making civilization possible. Fermented beverages can be made from a great variety of plants, and before civilization appeared man had probably already discovered that certain grasses serve to make excellent beer. Barley very early came to prominence in this regard, a position it has never relinquished. Today, of course, the grasses provide major ingredients for popular alcoholic beverages—rice is necessary for the making of sake, maize for bourbon, and sugar cane for rum. Whether this use of grasses has been in the service or disservice of mankind is the subject of divided opinions.

Tropical grasses of tremendous importance are the bamboos, exceptional in the grass family for their size and woodiness. Although they are practically indispensable in parts of the tropical world, they are little appreciated in temperate areas where some people may recall that they were once used for the pole vault and, of course, they are still used for fishing rods. I know of no better way to tell of the importance of the bamboos than to quote from a paper that a Thai student, Sa-korn Trinandwan, once wrote for my course in Economic Botany.

> I write this paper with gratitude and great respect for bamboo. In the poor families of the tropical people, the child is born on the floor of bamboo ribs under the bamboo roof in the bamboo hut. He is rocked in the bamboo cradle by his mother. He plays with toys which are made of bamboo. When he grows up and does something wrong, he will be punished by his mother with a bamboo rod. Sometimes he entertains himself with the bamboo flute. Sometimes he eats the bamboo shoots which are cooked on the bamboo fire. Then he grows up to be a man and makes love under the bamboo shade and builds a bamboo hut for his wife. When he grows old and dies, his body is buried in the bamboo coffin. The bamboos give sadness and happiness to the life of man from generation to generation. Bamboos are parts of men's lives. Bamboos are a part of their blood, and they seek their ways to the human soul.
>
> Bamboos can get along very well among the rich people. They get into the rich people's houses and serve them in the form of beautiful furniture. Sometimes bamboos get into their gardens and spend their entire lives with them. In the ancient times, even the king had to spend the night in the bamboo camp. During World War II, the prisoners of war spent part of their lives in the Japanese bamboo camps. How much the bamboos serve the man, whether rich or poor, high or low.

The importance of grasses for the feeding of livestock is fairly obvious. Although the saying that "all meat is grass" is not true, it comes close to being so since the grain and the leaves, either fresh or dried, of grasses are the principal food of most of our domesticated animals.

72

Figure 5-3
Some products of bamboo in Thailand. (Courtesy of Sa-korn Trinandwan.)

Not to be overlooked is the importance of grasses as ornamentals. Although a few grasses are occasionally grown singly for their aesthetic value, the chief claim to significance of grasses as "ornamentals" comes in their use for lawns. With the widespread movement to suburbia in the United States in recent years, billions of dollars have been spent for blue grass seed, or sod, or for other lawn grasses, plus fertilizers, mowers, and so on. At the same time a tremendous amount of money has been spent trying to eliminate undesirable plants from lawns, chief among them another member of the same family, crab grass.

In addition to crab grass, the family furnishes us with a number of other weeds, but then this is equally true of most large plant families. There is no good scientific definition of a weed. They are plants that follow man, growing best in the areas disturbed by him or his domesticated animals. They are sometimes referred to as "unwanted plants" or as "plants whose virtues have not yet been discovered." The latter definition has more than a modicum of truth in it. Rye and oats in prehistoric times were weeds of the cultivated wheat and barley fields. They spread with the domesticated plants as they were brought into new areas, and in northern Europe they grew better than the wheat and barley and in time became intentionally cultivated and eventually domesticated.

Finally, in this brief introduction to the significance of grasses to man, mention must be made of their importance to soil conservation. A cover of grass affords greater protection against erosion of soil than any other plant, for the blades bend to cover the ground while rain is falling, forming a mat that permits the water to run off with very little or no loss of soil. Various grasses are thus often planted in areas subject to soil erosion. Some grasses too, particularly rye, are planted to be plowed under to improve the texture and fertility of the soil.

Man has used grasses in many other ways, but rather than explore these, let us consider at greater length those that have contributed the most to man's food.

Wheat

The most widely cultivated plant in the world today, wheat, as was pointed out in an earlier chapter, was one of man's first two cultivated plants. The other, barley, has continued to the present day to be of importance, but chiefly as animal feed and as the source of malt for making beer. Wheat has become man's principal cereal, being more widely used for the making of bread than any other cereal because of the quality and quantity of its gluten. As it is gluten that makes bread dough stick together and gives it the ability to retain gas, the higher the proportion of gluten in the flour, the better for making bread.

Bread, of course, was not one of man's first prepared foods, for the primitive wheats first used by man were hardly suited for making bread. It seems likely that wild wheat and the early cultivated wheats were prepared by parching, which would have the advantage of removing the chaffy bracts surrounding the grain as well as making it more readily digestible. By grinding the parched grain and adding water, a gruel could be made. A beer of sorts may have been another early product in which grain was used. For a beer to be made it would be necessary for some of the starch* of the grain to be converted into sugar, which then could be fermented by wild yeasts. Man found such a brew to his liking, and although he may not have known it, it was a nutritious drink. Yeast, in fact, may have become a cultivated plant before the grains were, for residues from beer originally made from wild ingredients including yeast might have been used for starting new batches, meaning that the yeast was actually being cultivated. Some time later leavened bread resulted from the same process, the carbon dioxide bubbles formed during the fermentation process becoming trapped in the sticky dough and causing it to rise. Unleavened bread, of course, was probably used earlier and continues to have significance in certain religious ceremonies.

Wild wheats, which still are found in the Near East in some abundance, were being collected by man long before domestication occurred. Flint blades, which were apparently used for sickles to harvest grain, have been found in archaeological deposits dating back some 12,000 years. Milling stones and querns for grinding are even older, although we do not necessarily know that they were used for cereals. Evidence that the wild cereals could have supplied man with an abundance of food was provided by Jack Harlan, an American botanist, a few years ago. With a flint sickle he was able to harvest four pounds of grain in an hour. Thus, as he points out, primitive man could have in a space of a few weeks harvested more than enough grain to feed a family for a year.

We may never know for certain why the grain became cultivated but we do now have considerable information about how the wild plant became transformed into man's most important domesticated plant. We shall see that the process involved a number of accidental,

---

*Starch is hydrolyzed to a fermentable sugar by the action of certain enzymes. Such enzymes for making beer could have been supplied by certain molds, by a malt (that is, by germinating seeds), or by mastication. Malting barley is known to be quite ancient, being recorded on early written documents.

or natural, hybridizations followed by a doubling of chromosomes in the prehistoric period. The working out of the details of the origin of the domesticated wheats is as fascinating as any detective story (at least to the botanist) and the understanding of the origin of the plants has furnished knowledge of great importance to the plant breeder in his efforts to improve them.

Our present understanding of these origins is not the work of one or a few people, but results from a great many botanical studies conducted during this century in Germany, Russia, Japan, the United States, England, and Israel. Taxonomists, of course, had described many different species of wheat during the previous 150 years, and Linnaeus had provided the genus name, choosing *Triticum*, an old Latin name for cereal. In the early part of this century taxonomists recognized on the basis of the appearance of the plants that there were three groups of species of wheat. Shortly after, it was shown that the three different groups were characterized by different chromosome numbers, the diploids having 14 chromosomes, the tetraploids having 28, and the hexaploids, 42. Thus the wheat species formed a polyploid series with a base chromosome number of seven.*

The species with 14 chromosomes include the wild einkorn, *Triticum boeoticum*, which still grows wild in the Near East and the cultivated einkorn, *Triticum monococcum*, which differs little from its wild counterpart except in having slightly larger grains and less brittle rachises of the fruit stalks that prevent the grains from falling quite so readily. Einkorn is a low-yielding species, but is cultivated to a limited extent in the Near East and in central and southeastern Europe. The chromosome sets of the diploid wheat eventually came to be designated AA.

---

*The eggs and sperms of an organism each contain one set of chromosomes and are called haploid. The union of sperm and egg as a result of fertilization gives rise to a plant or animal designated as diploid that has two sets of chromosomes, one derived from each parent. Most plants and animals are diploid, but particularly in plants, we may find individuals or species characterized by having more than two sets of chromosomes and they are known as polyploids. Polyploidy results from an "accident" in the chromosome division in a species or more frequently in a hybrid between species. In the wheats there are two sorts of polyploids, tetraploids having four sets of chromosomes, and hexaploids having six. Chromosomes of a plant are studied by observation of dividing cells. Usually anthers in which young pollen grains are developing or young roots are used. The structures are crushed and treated with a stain that makes the chromosomes visible and then examined under a compound microscope at high magnification.

The 28-chromosome wheats comprise several species. One of these, *Triticum dicoccoides,* or wild emmer, is a wild species. The cultivated emmer, *Triticum dicoccum,* is grown principally in Asia and the Mediterranean area. Formerly used for making bread and pastries, it is now used mainly for livestock feed. Both of the emmers have covered, or hulled, grains, a characteristic of wild grasses. Several of the tetraploid wheats have naked grains that thresh free, a boon to man, of course, but detrimental for a wild species. Among the species with naked grains we find durum, or macaroni, wheat, *Triticum durum,* which is one of the important wheats today. The chromosome composition of the tetraploids is designated AABB, with each letter representing a set of chromosomes.

Several species with 42 chromosomes are also known, including some with hulled grains, such as spelt, *Triticum spelta,* once the principal wheat of Europe, and some with naked grains, such as bread, or common, wheat, *Triticum aestivum,* which has become the type most widely grown throughout the world today and preferred for bread. The hexaploids are designated AABBDD.

In a polyploid series the most ancient species is the one with the lowest chromosome number, and from this it would follow that einkorn (AA) is the starting point for the derivation of the other wheats. Einkorn hybridized with another species with chromosome sets BB and following chromosome doubling gave rise to the tetraploid, or AABB, species. One of these tetraploids, in turn, hybridized with still another species with the DD chromosome constitution and this triploid hybrid in turn doubled its chromosomes to give rise to the hexaploid, or AABBDD, wheats (Figure 5-4). Through a series of brilliant investigations, the other species involved in the origin of the tetraploids and hexaploids have been fairly certainly identified. Both are species of the genus *Aegilops,* or goat-grass,* a weedy grass worthless to man except for its contribution of chromosomes to wheat. The contributor of the BB chromosomes is probably *Aegilops speltoides,* which is found from southeastern Turkey across the Fertile Crescent, or a species similar to it. The DD chromosomes, which contributed genes for the high gluten content of the hexaploids as well as making them better adapted to extreme environments than are the other wheats, come from *Aegilops squarrosa,* a species that

---

*Some people now place the goat-grasses in the same genus as wheat *(Triticum)* instead of in *Aegilops.*

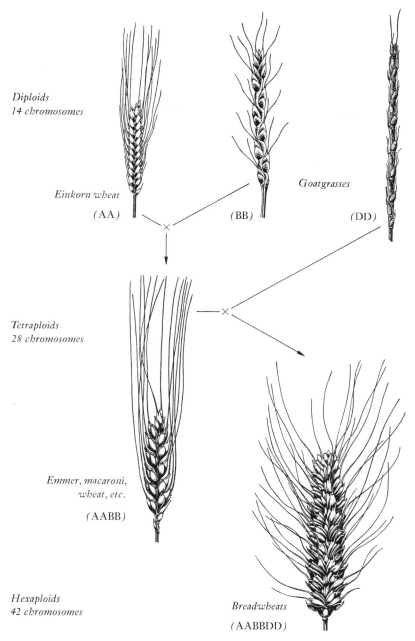

Diploids
14 chromosomes

Einkorn wheat

(AA)

Goatgrasses

(BB)

(DD)

Tetraploids
28 chromosomes

Emmer, macaroni,
wheat, etc.

(AABB)

Hexaploids
42 chromosomes

Breadwheats

(AABBDD)

Figure 5-4
Evolution of domesticated wheats. (From "Wheat" by Paul C. Mangelsdorf.
Copyright © 1953 by Scientific American, Inc. All rights reserved.)

ranges from eastern Turkey to Kashmir and Pakistan. Man apparently had no role in the development of the polyploid wheats except perhaps for unwittingly bringing the cultivated plants and weeds together so that it was possible for hybridization to occur. In addition to the accidental hybridizations and chromosome doublings that gave rise to the tetraploid and hexaploid wheats, one other event was of paramount importance in their evolution. This was a mutation that influenced the chromosome pairing in the polyploid plants so that they had good fertility. Ralph Riley, who with V. Chapman is responsible for our knowledge of it, has called attention to the importance of this single mutation for the development of the whole of Western civilization.

Archaeological work has helped to supply approximate dates as well as to indicate the general areas where the wheat species developed. Wild emmer has been reported from the archaeological site of Tell Mureybat on the banks of the Euphrates dated at about 8000 BC. Wild einkorn, einkorn, and emmer are all found at Ali Kosh in southwestern Iran in deposits dated between 7500–6750 BC, and both wild einkorn and emmer occur at Haçilar dated at about 7000 BC in west central Anatolia. At Jarmo in the Iraqi Kurdistan in deposits dated at about 6750 BC, both wild and cultivated einkorn have been found along with grains that appear to be somewhat transitional between wild and cultivated emmer. Deposits in Greece dated back to 6100 BC reveal both einkorn and emmer. From these finds it appears that man may still have been collecting wild wheats at the same time cultivation was beginning, which is hardly surprising since the early cultivated plants were probably neither high yielding nor extensively grown. Both einkorn and emmer appear to have come into cultivation at about the same time, with emmer being the more widely cultivated. Bread wheat, as might be expected, came somewhat later. There is archaeological evidence for its presence at Çatal Hüyük in Anatolia in the period 5850–5600 BC, at Tepe Sabz in southwestern Iran between 5500–5000 BC, and at Haçilar between 5800–5000 BC. A tentative identification would place the species in Greece at about 6000 BC. The times of the appearance of the various wheat species will probably have to be revised somewhat as future archaeological work takes place.

Although the earliest cultivation was apparently confined to the Near East, the wheats were soon to become widespread. Emmer was

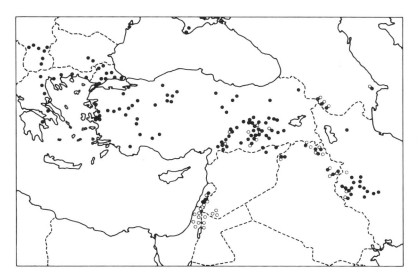

Figure 5-5
Distribution of wild einkorn (solid dots) and wild emmer (open circles).
(Adapted from "Distribution of Wild-Wheats and Barley" by Jack R. Harlan
and Daniel Zohary, *Science*, 153: 1074–1080, 1960. Copyright © 1966 by the
American Association for the Advancement of Science.)

to become the principal wheat of ancient Egypt.* From early cultivation in the Balkans wheat spread to other parts of Europe. One of
the hexaploids, spelt, may have originated in central Europe, where
it was prominent in prehistoric and early historical times. Other
hexaploid wheats appeared in India, probably before the fourth millenium BC, and in China sometime before the beginning of the Christian era. Some people think mainly of rice in connection with both of
these countries, but wheat early became an important plant there and

---

*Occasionally an item is published in a newspaper to the effect that Mr. and
Mrs. So and So on a vacation to Egypt visited the pyramids and brought home
with them a few grains of "mummy" wheat, several thousand years old, and
that the grains were planted and gave rise to healthy wheat plants. While it is
true, of course, that grains of wheat dating back to pre-Christian times have
been found in Egypt, there is no scientific verification of their ever germinating.
Wheat seeds, as is true of most plants, have a rather short life lasting only a few
years.

continues to be one in the more temperate parts of the countries. The Spanish brought wheat to the Americas, and eventually the United States, Canada, and Argentina were to become among the world's greatest producers. (As token of the way technological and scientific development is drastically altering agricultural practice, it is interesting that the United States, where wheat is a relatively new crop, recently sent agricultural specialists to Turkey to give advice on growing wheat, where the crop has been grown for thousands of years.)

Wheat grows best in cool climates with little to moderate amounts of rain, and many areas of the earth's surface are suitable for its cultivation. Wheat is being harvested somewhere in the world every month of the year. Of the several species brought into domestication, only two are of major importance today. *Triticum aestivum,* used primarily for flour for bread and pastries, and *Triticum durum,* principally employed for making paste products such as macaroni, spaghetti, and noodles. Both species are cultivated in the United States, the latter principally in North Dakota. In the northern part of the United States and in Canada, wheat is planted in the spring; to the south of this area it is sown in the fall.

Although in some parts of the world the methods employed in growing, harvesting, and utilizing wheat are only little changed from those used in prehistoric times, in other parts there have been great advances in wheat farming and processing the grain—from the introduction of modern mechanical planting and harvesting to steam-driven rollers for milling. In the first part of this century it still took tremendous manpower to thresh and harvest wheat although some machinery was used. The combines used today consolidate what were formerly many separate operations of threshing and harvesting, and can be operated by a single person. The combine has thus led to drastic changes in farm life, including the virtual disappearance of the old threshing "party." Forty years ago straw stacks were a common sight on farms in the midwestern United States—now the straw is usually scattered on the fields in the combining process or neatly baled. The straw, although of minor importance compared to the grain, has some uses. Not a nutritious hay for livestock, wheat straw is unusually strong and has many uses on a farm and can be used also by manufacturers as a filling for mattresses and in the making of paper and paper board.

Man's bread, too, has changed. To make a white flour it is necessary to remove the germ and the bran from the grain. Technological advances in the milling process over the years have led to a whiter flour. Since the natural product is somewhat yellowish in color a bleach is sometimes employed, and chemicals, such as sodium propionate, are sometimes added during the baking process to retard spoilage of the bread. Although white flours have superior keeping and baking qualities, they are obviously much less nutritious than the whole wheat. Therefore, the flour is often enriched by the addition of vitamins and minerals which is required by law in some countries. These, however, fail to compensate for all the nutrients that have been removed from the grain. In the past a white bread was associated with the "higher" classes or, at least, those with money. Recently there has been a trend toward other kinds of bread, at least among some of the better educated. It is a sad note on modern civilization that man must remove some of the nutrients and then supply additives to the product that goes on the table.

There have been changes in the wheat plant itself as well as in methods of growing, harvesting, and milling it. The evolution began, of course, at the time man began cultivating the wild grasses. As the wild plant had a brittle fruiting stalk, the grains tended to fall individually as they matured, an advantage for seed dispersal in nature but undesirable for man, who wanted to collect all the grains at one time. With domestication the fruiting stalk became less brittle, permitting the grains to remain on the plant. The cultivated plants developed seeds that would germinate evenly and rapidly in contrast to the slower and irregular germination of those of the wild plant. The bracts surrounding the grain became more loosely attached, letting the grains fall free during the harvest, which made it easier to prepare them for eating. There was also some increase in the size of the grain. Numerous varieties of the different species came into existence through the agencies of mutation, hybridization, and both conscious and unconscious selection. Possibilities for hybridization increased as man moved from place to place or exchanged seeds with other people.

As new varieties of wheat come into existence, some of the older ones tend to disappear, but many exist little changed for hundreds of years. In this century plant breeders through conscious selection and artificial hybridizations have produced so many new varieties that today there are probably more than a thousand different kinds of bread wheat alone, some of them tailor-made for high productivity,

A

B

C

D

Figure 5-6
Traditional and modern methods of harvesting wheat. *A.* Harvesting one of the
new high yielding wheats in India. (Courtesy of FAO.) *B.* Threshing wheat
with oxen in India. (Courtesy of Rockefeller Foundation.) *C.* Milling wheat in
Pakistan. (Courtesy of FAO.) *D.* Combining wheat in the United States.
(Courtesy of USDA.)

special milling properties, adaptation to different climatic conditions, and resistance to disease.

Perhaps it is resistance to disease that has occupied as much of the plant breeder's attention as any other character. Practically all of man's cultivated plants are subject to a number of diseases, and the wheat plant is particularly susceptible to stem rust, caused by a fungus that belongs to the same group as the common mushroom but is profoundly different in appearance. A severe infestation of stem rust can almost completely destroy a crop. Many races of wheat have been bred that are resistant to rusts, but the great difficulty is that there are also numerous races of rusts and new, more virulent ones continually come into existence through mutation and hybridization. Thus there will probably always be a need for the plant breeder to attempt to develop new resistant strains of wheat. Another accomplishment of the plant breeder has been his contribution to the doubling of the wheat yield in the United States in the course of little more than a quarter of a century. Greater use of fertilizers and improved cultural practices have also contributed to the increased yields.

Among the recent achievements of the wheat plant breeder and other agricultural scientists is the remarkable work accomplished in Mexico that has contributed significantly to the green revolution. In 1943 the Rockefeller Foundation, at the invitation of the Mexican government, went to that country to see what could be done to increase food production. Although maize is the basic food plant of Mexico, wheat is very important also and Mexico was then importing half of the wheat it consumed. Much of the wheat being grown in Mexico at the time was little changed from that originally introduced by the Spanish in the early part of the sixteenth century. Working with Mexican agronomists, the scientists from the Rockefeller Foundation were able to double the yields in 20 years. By 1956 Mexico was self sufficient in wheat and since that time has produced some for export. A large part of the success resulted from the development through hybridization of rust-resistant varieties and dwarf forms able to take heavy applications of fertilizer without falling over. The older, taller varieties tended to lodge, or fall over, when given sufficient fertilizer to increase the yields and thus could not be readily harvested. The parental type of the new dwarf forms that have proved to be so successful originally came from the Orient. The first crop planted by Norman E. Borlaug, who headed the work for the

Figure 5-7
Dr. Norman Borlaug (fourth from left) showing one of the new wheats
developed in Mexico to a group of visiting scientists. (Courtesy of the
Rockefeller Foundation.)

Rockefeller Foundation and in 1970 received the Nobel Prize for his
contributions to agriculture, was lost to rust. Fortunately a few of
the original seeds were still left in the seed envelopes and these were
grown the next year and used for the crosses that ultimately gave
the kinds of plants that were wanted. Some of the varieties developed
for Mexico have proved to be successful in Pakistan and India, and
Pakistan is now producing enough wheat for its own needs while
India is approaching that goal. It is reported that in parts of India
that used to give 7–8 bushels an acre, yields of up to 60–70 bushels an
acre are being secured by growing the new varieties under irrigation.
In India the new high-yielding varieties were attracting such attention that armed guards had to be posted at the experimental stations
to prevent people from stealing seed before it was ready to be released to the public. Since wheat is a self-pollinated crop, once a
hybrid has been stabilized and released to a farmer, he can multiply
his own seed stock for future planting.

Some years ago the Russians made great claims for a perennial
wheat that could be grown from year to year without replanting.
Through hybridization of a wheat with a perennial grass such a plant

Figure 5-8
Armed guard protecting plot of new strain of wheat at experimental station in India. It was necessary to post guards so that the farmers would not take the seeds before they were ready to be released to them. (Courtesy of Rockefeller Foundation.)

was actually made but thus far it has been of little significance. Of potentially greater significance, perhaps, is a man-made hybrid between wheat and rye first produced in Sweden. Although originally almost sterile, fertility has been built up through selection. The hybrid, called *Triticale* (from *Triticum*, wheat, and *Secale*, rye), gives grain that contains more and better protein than wheat, but as it is low in gluten, it cannot be used for making bread. Research is continuing in some places, particularly in Canada, on this "man-made" cereal plant.

One of the next great "break-throughs" in wheat breeding may come with the production of a hybrid wheat that cashes in on the phenomenon of hybrid vigor—the increased yield of first-generation hybrids—as does hybrid maize, which is discussed later in this chapter. Hybridization, both natural and man-made, has been of importance in the development of the wheats but is far more difficult than in maize, owing to the differences in their flowers: Although both are grasses, wheat has bisexual flowers, whereas the sexes are separate on corn plants, making the production of hybrids a relatively simple matter.

Although it is always important to produce a plant that can yield more per acre, production of wheat was so great in the United States in the early 1960's that it included what has been referred to as an "embarrassing" surplus, more than could be utilized at home or sold for export. It has been easier to breed high-yielding wheats than it has been to solve the problems of their distribution.

Rice

In Japan each year the Emperor himself, still patron of all agriculture although no longer regarded as the descendant of the sun, joins in the ritual harvest of the rice on the small imperial paddy field in the palace grounds. Rice is still regarded as a sacred plant by many people in much of Asia. Although of more limited distribution than wheat, rice feeds more people, since it constitutes the basic food of more than half of mankind.

Far less is known about the origin of rice than about the other important cereals, but there is little doubt that its cultivation arose in southeastern Asia where it continues to be of paramount importance. Archaeological evidence of it is known from Thailand dating back to the period from 5000 to 3500 BC, but it could be that its origin is even earlier, for the archaeological record of plants from this part of the world is much poorer than it is for the Near East and the Americas.

Rice reached Japan from China during the second century BC. At this time it was already known in Europe, for it had been carried to Greece by Arab traders and again during Alexander the Great's invasion of the East. The scientific name *Oryza*, applied to the plant by Linnaeus, comes from an ancient Greek word for rice that, in turn, is derived from an Arab word or a Chinese word—opinion is divided among scholars. The Chinese word means "good grain of life"; not surprisingly, the word for rice is the same as that for life, food, or agriculture in many parts of the Far East. Rice is mentioned in very ancient Chinese writings, as it is in early Hindu scriptures. Several

88

Figure 5-9
Rice paddies. (Courtesy of FAO.)

different varieties of rice are described in the ancient Dravidian literature of India. Linguistic evidence, as well as botanical, suggests an origin for the plant somewhere in southeastern Asia.

There are about twenty-five species of *Oryza* and of these, one species, *Oryza sativa*, furnishes virtually all of the rice of the world today. A second domesticated species, *Oryza glaberrima*, is still cultivated in West Africa, presumably in the same area where it originated. Of the wild species, *Oryza perennis*, widely distributed throughout the humid tropics of the world, appears to be the most likely progenitor of *Oryza sativa*. This wild rice* presumably was collected by primitive man, just as wheat was, and eventually became intentionally grown. With cultivation, mutant types were selected and in time man had an annual, with larger grains and nonshattering fruit stalks, and sometimes lacking the awns of the wild species. Hybridization of the cultivated rice with various wild species also appears to have contributed to the development and the great variability of this plant. Wild rice is still collected by some people, and is the preferred plant in some ancient religious ceremonies in southeastern Asia. Although some tetraploid races are known, most varieties of rice are diploid, having 24 chromosomes.

Rice was introduced into the Carolinas in 1647, but today California, Arkansas, Louisiana, and Texas are the rice-growing areas of the United States, and more than half of the rice produced goes for export. As the average per-person consumption of rice in the United States is only six pounds a year, not a great deal need be produced to meet the domestic demand. Some rice is grown in Africa, South America, and Europe, and it is an important food in many areas. In many parts of Latin America, rice and beans are served with every meal and sometimes constitute the whole meal. But it is in its original homeland that rice is dominant, where more than 90 percent of the world's crop is produced, and where people on the average consume as much rice in a week as the average American eats in a whole year.

Throughout much of southeastern Asia, rice is grown as it has been for many centuries, requiring tremendous human effort. It has been estimated that in some regions one thousand man hours are needed to grow and harvest a single acre of rice. A great many people still cultivate very small patches of rice, from one to five acres.

---

*There is another plant known as wild rice that is also used for food; it is a grass but is not very closely related to true rice, or *Oryza*. This wild rice, *Zizania aquatica*, is an annual, native to eastern North America, and was an important food plant of the Indians of the Great Lakes region, particularly in what is now Minnesota, Wisconsin, and Manitoba. The plant is not cultivated but it is still collected by Indians for their own use and as a cash crop. At one point outsiders introduced mechanical means of harvesting the crop that could have proved disastrous if their use had gone unchecked. The more primitive methods always left plenty of seeds to re-establish the plants for the next season, but with the more efficient mechanical method, most of the seeds were collected and the species was threatened with extinction. Fortunately some laws were passed governing the harvest in time that the plant still survives.

Figure 5-10
Terraced paddy fields in Indonesia. Mechanization would be difficult in this kind of terrain. (Courtesy of FAO.)

A

B

Figure 5-11
A. Farmer hoeing a paddy field in Indonesia. Although water buffalos (Figure 4-8) are widely used, much hand labor still goes into the preparation and care of the fields.
B. Transplanting rice seedlings in Japan. (Courtesy of FAO.)

As methods vary somewhat from region to region, a generalized description is impossible, but the following steps are very common. The paddy fields are prepared for planting with a wooden, single furrow plow drawn by a water buffalo or ox. Manure, if available, is scattered on the field. The plowed land is then smoothed with a log. After the dikes have been repaired, river water drawn by primitive water wheels is used to flood the fields in areas where rain is not sufficient for this purpose.

The rice is then planted, either by broadcasting dry or previously germinated seed or by transplanting seedlings or young plants that have been grown in a nursery bed. The customary way of planting most cereals is directly by seed, and knowledge of the origin of transplanting seedlings by hand would be of interest. Carl Sauer has postulated that the latter is the older method, first used by people who had practiced vegetative cultivation of other crops, transplanting being similar to starting plants from cuttings or by other vegetative means. Broadcasting dry seeds might have been derived from contact with wheat farmers at a later date. It is not definitely known which is the older method. Transplanting, backbreaking work that requires much stooping, is usually done by women and children. Some pruning of the tops and the roots of plants normally accompanies transplantation, and some people believe that this pruning stimulates growth of the young plant and leads to higher yields than when seeds are broadcast. Weeds present less of a problem when rice is transplanted than when seed is planted directly, and having less competition from weeds could also help account for greater yields being obtained through the transplanting method. After the plants are established, any necessary weeding is done by hand, another activity that requires much stooping.

Harvesting is done with sickles or knives. With the latter the seed heads are gently cut individually, the idea still existing in some regions that if the cutting is harsh, the rice plant will be offended and the yield will be decreased the following season. Threshing is done by beating the heads against the ground or against logs, or by having animals or barefoot humans tread upon the seed heads. Sometimes women who do the treading are barebreasted, which is thought to be related to an ancient belief that the less they wear, the thinner the rice husks will be. Winnowing to remove the chaff from the grain is still often accomplished by tossing the rice from bamboo or rattan trays and allowing the wind to blow away the lighter chaff while the

Figure 5-12
Harvesting rice in Indonesia. (Courtesy of FAO.)

grain settles nearby. All of these methods, of course, contrast strikingly with the highly mechanized procedures practiced in the United States and some other countries, where airplanes are sometimes employed for seeding, which allows the grain to be produced economically in spite of high labor costs.

Following winnowing, under primitive conditions a mortar and pestle are used for hulling and the brown rice that results may be promptly cooked for local consumption. In industrial preparation pearling or whitening is used to remove the bran, or outer layers of the grain, and sometimes this is followed by polishing. The final product is pleasing in appearance and taste but less nutritious than brown rice. Unfortunately, just as with bread, most people prefer white rice to brown. The loss of nutrients, particularly vitamin B_1, eliminated in the process of milling and cooking, has been responsible for much malnutrition and diseases, such as beri-beri, among people whose diet consists almost entirely of white rice. Highly milled rice contains only 0.04 mg. of vitamin B_1 per 100 grams compared to 0.40 mg. in the unmilled; the latter amount is sufficient to prevent beri-beri. White rice, of course, can be fortified with additives to increase its nutritional value.

Rice, although not a true aquatic plant, is unusual among the cereals in that its roots can thrive under water. Thus it is an ideal plant for much of the humid tropics, although at times it can get too much water, for proper drainage is essential to good growth. Rice also can be grown much as the other cereals. Dry, upland rice, or hill paddy, is grown in some areas that have the proper temperature and sufficient rainfall. Yields, however, are generally lower than for lowland rice.

The thousands of "varieties" of rice are generally classed or divided into two major groups—the *japonica* types, which have short grains and are sticky when cooked, and the *indica* types, which have long grains and are drier when cooked. The *japonica* types, grown in both Japan and Taiwan, in general are higher yielding than are the *indica* types, which are grown throughout the greater part of Southeast Asia. World hunger could be at least partially reduced through the wider distribution of the higher yielding varieties. One of the obstacles to achieving this goal, however, lies in the difficulty of persuading people to accept new foods. People the world over usually prefer their local varieties and if they are used to eating a sticky rice, for example, they are often reluctant to change to a dry type, and vice versa.

Figure 5-13
A. Winnowing rice in Burma. Courtesy of FAO.)
B. Hulling and pounding maize in Cambodia. Many of the traditional
implements used for rice have been adopted for maize in this part of the world.
Banana plants in background. (Courtesy of FAO.)

A

B

The rice plant, of course, directly or indirectly serves man in many ways other than for food. Beers and wines may be made from the grain, the most famous of which is sake, the "national beverage" of Japan. Rice is also used in the manufacture of beer in the United States and other countries. Rice is used for the production of starch, and rice powder is used as a cosmetic in parts of the Far East. The hulls or husks of the grain are used as fuel, in making building materials, for the manufacture of furfural, which in turn is used to make plastics, and in other ways. The straw, of course, is not overlooked and is used in the manufacture of baskets, mats, and strawboard. Like wheat, the straw is not very nutritious but is sometimes used for livestock feed. Paper can be made from rice straw, but what is commonly known as rice paper is made from another plant—*Tetrapanax papyriferus*, or rice-paper plant—which is native to the Orient but is not a grass.

Perhaps the most important by-product of rice farming doesn't come from the plant itself, for paddy fields are frequently used for raising fish. A rice diet is extremely deficient in protein, and fish, of course, is an excellent protein source. In many rice-growing areas ponds are kept for fish hatcheries, and minnows, usually carp, are introduced into the paddy fields. The fish, rather than decreasing rice production, actually appear to promote its fertility. Without the fish in the wet rice fields, other problems may arise—as was discovered in the south coastal area of the United States some years ago when huge swarms of mosquitoes found the rice fields ideal breeding grounds. In parts of Asia the recent use of modern pesticides has had a deleterious effect on the fish populations.

Of the many diseases of rice, one has proven to be of particular scientific interest. The *bakanae* or "foolish seedling" disease, which produces unusually tall, thin plants, was found by Japanese botanists to be caused by a fungus, *Gibberella fujikuroi*. They found that a growth substance, now known as gibberellin, could be isolated from the fungus, and this substance has been the subject of considerable scientific work by American and English botanists as well as the Japanese since World War II. Gibberellins are used to produce growth in some dwarf plants and to induce flowering in others.

Although some efforts to improve the rice plant have been carried on for many years and high-yielding strains have been grown in Japan and Formosa, much of Asia still grows its old unproductive types under very primitive conditions. Recognizing the need for improvement in a plant that feeds a great part of the world's population, the Ford Foundation in cooperation with the Rockefeller Foundation established the International Rice Research Institution in the Philippines. With scientists representing many different disciplines assembled from the United States and six eastern nations, the IRRI began its work in 1962. It was soon found that although rice was a much studied plant there was still much room for basic research. Effort was devoted to learning more about the plant and this search for basic knowledge still continues. It was known that one of the great drawbacks of Asiatic rice was its tendency to lodge, or fall over, as the plants reached maturity. Fertilization that would improve yields of such plants would only contribute to lodging since it would produce taller plants with heavier heads. The rice scientists realized that an ideal rice plant would be a short one with a strong stem that could take additional amounts of fertilizer. Over 10,000 different samples of rice seed from many different areas were assembled. Through hybridization, the IRRI scientists attempted to produce a plant having the desirable characters, and one of the crosses, involving a short rice from Taiwan called Deegeowoogen and one from Indonesia called Peta, has already shown extremely high yields where it has been tested, not only in the Far East but in South America and Africa as well. The new plant, designated IR8, reaches maturity in 130 days and is insensitive to day length, making it adaptable to many regions and making it possible to grow more than one crop a year in some places. Other high-yielding varieties have since been developed. The IRRI did not neglect other aspects of growing rice—fertilization, irrigation, disease control—and even produced an inexpensive threshing machine that could be widely used. Thus in the space of a few years through a highly concentrated effort rice yields increased dramatically, which is already contributing to the green revolution. What is still needed is a rice that has a higher protein content, for rice contains even less protein than wheat and maize, and the newly produced varieties may have protein contents even lower than that of some of the older, low-yielding varieties.

Maize, or Indian Corn

From a plant that was once used mostly to feed people, maize has become one of the world's chief feeds for animals. The plant, however, still retains its role as a basic human food plant in parts of the Americas. It is generally held that maize was not known in the Old World until after some of Columbus' men found it growing in Cuba. Maize was already several thousands of years old at the time, and it had fed the laborers who built magnificent temples in Mexico and Peru in prehistoric times; today, directly or indirectly, it supplies much of the energy for the technological developments in the United States.

At the time of the Discovery, maize was the most widely grown plant in the Americas, extending from southern Canada to southern South America, growing at sea level in some places and at elevations higher than 11,000 feet in others. Woodlands were cleared, swamps drained, deserts irrigated, and terraces built on mountain sides so that the plant could be grown, but little maize grew in the area that was to become the great corn belt of the United States, for not until the mould board plow was invented would it be possible to turn the heavy prairie sod. For many years after America was "discovered" much of the future corn belt was to continue to be dominated by wild grasses and buffalos.

Much of American Indian life centered on this "gift of the gods," just as it still does today in parts of Mexico and other Latin American countries. In religion and art as well as in everyday life many of man's activities were concerned with the maize plant. For food, it was popped, probably one of the most primitive ways for man to prepare the hard grain, parched, boiled, or ground. By washing the grains in wood ashes and quicklime, the Indians made hominy of it. Ground, it was used for making "bread," or tortillas. It was also eaten green, and from quids, or chewed wads, recovered in archaeological deposits, we know that this is one of the most ancient methods of eating it. Sometimes even the pollen was added to soups or stews, and corn smut, a fungus disease that affects corn, was also eaten. As Paul Weatherwax has written, "Perhaps the most cheering and heart-warming use the Indians made of maize was the production of alcoholic beverages." It was probably early learned that a nutritious beverage could be made by chewing or germinating the grain to start a process of fermentation. The beer, or *chicha*, that results is still widely made and used in South America, although unfortunately it has been replaced by the much more potent and less nutritious sugar cane alcoholic beverage, aguardiente, in many areas. The Indians also found many uses for parts of the plant other than the grain, as they still do.

Figure 5-14
Indian methods of preparing bread and chicha.
(From Jerónimo Benzoni, 1565.)

In Europe it was first grown as a curiosity, like many other American plants of the time. To distinguish it from the other cereals it was at first called Turkey corn, or Turkey wheat because some people thought that it came from that country.* An American Indian name, maize, or mays, was used to some extent. Linnaeus, the great Swedish botanist of the eighteenth century who named many of our plants, adopted it for the specific designation of the plant and used *zea*, an ancient Greek word for cereal, as the genus name. Thus we still have *Zea mays* as the scientific name of the plant.

The new plant was not exactly a rousing success in all parts of Europe. John Gerard, the famous English herbalist of the late sixteenth and early seventeenth century, had this to say:

> Turky wheat doth nourish far lesse than either wheat, rie, barly, or otes. The bread which is made thereof is meanely white, without bran: it is hard and dry as Bisket is, and hath in it no clamminesse at all; for which cause it is of hard digestion, and yeeldeth to the body little or no nourishment. Wee have as yet not certaine proofe or experience concerning the vertues of this kinde of corne; although

---

*Some people have maintained that maize reached Asia before Columbus discovered America, and that it may have entered Europe by way of Turkey.

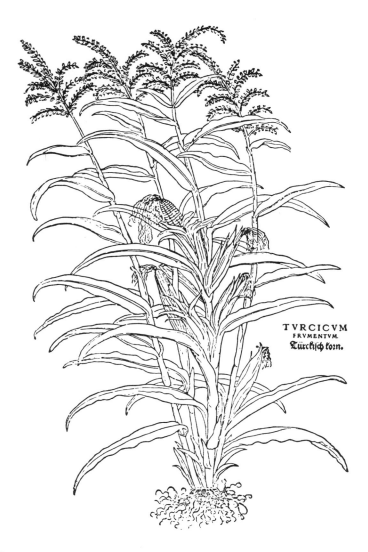

Figure 5-15
First published illustration of maize in Europe, from herbal of Leonhard Fuchs, 1542. Fuchs believed that the plant came from Asia and called it Turkish corn.

the barbarous Indians, which know no better, are constrained to make a vertue of necessitie, and thinke it a good food: whereas we may easily judge, that it nourisheth but little, and is of hard and evill digestion, a more convenient food for swine than for man.

More than two centuries later, the English imported corn to Ireland from America as food for the Irish people during the great famine. The Irish did not take readily to the new food. They did not know how to prepare it, they had no good implements for grinding it, and when they did, they didn't like the taste. Their inability to accept and use a New World plant is rather ironic, for it was the failure of the potato crop that led to the famine and the potato had been introduced to Europe from the Americas even later than maize.

Indeed, the maize plant was a strange one to the European—an amazing plant, if you will excuse an old pun—quite unlike the other cereals known to them. Not only was it larger, but instead of bearing its grains in a head at the top of the plant it bore them in ears on the sides of the stalk with the silks, or styles, protruding. At the top it produced tassels, which, except in exceptional circumstances, produced only male flowers. The grains, instead of being covered individually by chaff like the other cereals, were naked* and the ear as a whole was covered by husks. Although nearly all domesticated plants are poorly equipped to survive in the wild, maize is even more helpless than most, because the grains remain attached to the cob. Maize had to be handled differently for planting than the cereals familiar to the Europeans. Instead of being broadcast over the field, maize grains, which are much larger than those of other cereals, were planted individually. This method of planting may relate to a basic difference of very early agriculture in the Old and New Worlds. In the Old World man had animals to plow the fields in preparation for the broadcasting of seed, whereas in the New World the only animal available was man himself and what crude preparation the field received was done with a digging stick or hoe. The maize farmer's attention would have been more focused on the individual plant, for which he had carefully sown a single seed, than the wheat farmer's, who had broadcast seed; this may have been significant in the evolution of maize, for attention by man to variant individuals may help explain the great diversity found in maize.

---

*There is a rather rare form of maize, known as pod corn, in which the individual grains are covered.

TRAVAXO
3ARATARPVMITAN

TRAVAXA
3ARAPAPAHALLMAIMTA

C

D

Figure 5-16
Growing maize in Peru. *A*. Planting. Footplow is used to break soil.
B. Cultivating. (From Poma de Ayala, c. 1600.) *C*. Preparing a field for planting
today. Note similarity of plow to that shown in *A*. (Courtesy of FAO.)
D. Terraces. Many of the terraces constructed by the Incas are still in use at
present. (Courtesy of Paul Weatherwax.)

Maize is certainly a most variable species, probably more so than any other plant. Many people are well aware of the great variation in the color of the grains, since multicolored ears are often used as decorations. The kernels also vary considerably in size and shape as do the ears, which may range from a little over one inch to a foot and a half long. Some varieties are known that reach maturity in a little over two months whereas others require more than a year. For convenience, five main types of maize are recognized: (1) popcorn, probably the most primitive type, with extremely hard grains that allow pressure to be built up within them upon heating; "popping" results when the sudden expansion of the soft starch turns the grain inside out (other varieties of maize and other cereals, of course, can also be popped and are commonly prepared in this way as breakfast foods, but special methods are necessary to allow the pressure to be built up); (2) flint corns, which have kernels of hard starch; (3) flour corns, which have soft starch, of particular value to the Indians because it is easily ground, but disadvantageous in being quite subject to insect damage; (4) dent corns, so called because there is a dent in the top of the kernel in which a soft starch overlays an area of hard starch; (modern dent corns are responsible for the high productivity in the corn belt, and were originated in the nineteenth century by crossing a southern dent corn with a northern flint variety); (5) sweet corn, which has a sugary instead of starchy kernels, and is today a favorite fresh vegetable. There is also another type of corn, called waxy corn, that is distinguished by a starch that is chemically different from that of other corns. It receives its name from the wax-like appearance of the grain when it is cut. Waxy corn was discovered in China at about the beginning of this century, and has come to have some special uses in industry and for food because of its very different kind of starch. In spite of the great variability, all the corns are regarded by botanists as belonging to a single species. All of them are diploid, having ten pairs of chromosomes.

Where this species came from has long been a question of great interest to botanists, and at times proponents of different views have had rather heated arguments. Although much remains to be learned, there now appears to be some consensus of opinion about where maize must have originated and the approximate time when this occurred, as well as about what happened to the plant under man's influence. There also appears to be growing agreement about what plant gave rise to maize.

Figure 5-17A
A collection of maize from Guatemala showing some of the great variation
found in this plant. (Courtesy of USDA.)

Figure 5-17B
Varieties of corn. Left to right: popcorn, sweet corn, flour corn, flint corn, dent corn, pod corn. (Courtesy of USDA.)

It has been known since the last century that teosinte, a coarse, wild grass of Mexico, Guatemala, and Honduras, now becoming rather rare, was the closest relative of maize. The two plants cross naturally and produce fertile hybrids. Teosinte was originally considered to belong to a distinct genus and was classified as *Euchlaena mexicana*. The later recognition of its very close relationship with maize prompted its transfer to the genus *Zea*, and quite recently the proposal has been made that it belongs to the same species as maize. The idea that teosinte might be the ancestor of maize is not a new one but it was largely ignored for a number of years in favor of a much more spectacular hypothesis of the origin of maize.

The view that teosinte was not an ancestor of maize but a .hybrid that had maize as one of its parents was developed by Paul Mangelsdorf and R. G. Reeves. Their hypothesis, advanced in 1939, proposed that maize had originated in South America from a wild form of pod corn. It was introduced by man into Central America and Mexico, where it encountered the wild gamagrass *Tripsacum*. Hybridization took place between maize and *Tripsacum*, and, as a result of this hybridization, teosinte was created. Hybridization later took place between maize and teosinte, producing superior varieties of maize. In support of their hypothesis, they pointed out that pod corn, which had a natural means of dispersal for the grains, could be construed as the wild type; that the Andes could be regarded as the center of the origin because there the greatest variability* in maize is found; and that teosinte was in many respects intermediate between maize and *Tripsacum*, as might be expected of a hybrid.

Since 1939 considerable new archaeological evidence forced Mangelsdorf to revise his hypothesis about the place of origin. Pollen discovered under Mexico City, dated as being 80,000 years old, was identified as that of maize. A very primitive type of corn was discovered at Bat Cave, New Mexico, more than 4000 years old, far older than any maize known from South America. Still more evidence that maize originated in Mexico came from Tehuacan, where cobs of maize only a little more than a half inch long were found in deposits dated between 5200 and 3400 BC. At the time of their discovery it was postulated that these remains came from a wild maize that later became extinct.

Although a hybrid origin for teosinte received fairly wide acceptance for a number of years, recently more and more investigators have concluded that it is a "natural" species. No hybrids between maize and *Tripsacum* have ever been found in nature, and although it is possible for man to make the cross, special techniques are usually required to do so, and no teosinte-like plants have been produced from such crosses. If teosinte is the progenitor of maize there is, of course, no need to postulate that wild maize ever became extinct, for it and wild maize are the same. The pollen under Mexico City, if indeed it is 80,000 years old, would have to be that of teosinte, and the primitive maize found at Tehuacan would have to be regarded as an early domesticated form of maize.

One of the stumbling blocks in the acceptance of teosinte as the ancestor of maize had been that it was difficult to understand how

---

*The idea that the center of diversity of a cultivated plant usually indicates its place of origin was formulated by N. I. Vavilov, a Russian scientist who made many contributions to our understanding of domesticated plants. Today it is realized, however, that there are often other explanations for centers of diversity.

A

Figure 5-18
A. Teosinte. (Reprinted with permission of The Macmillan Company from Paul
Weatherwax, *Indian Corn in Old America*. Copyright © 1954 by The
Macmillan Company.) *B*. Teosinte ear on left, about three inches long.
Husks removed to show single row of grains on right. The grains are separate at
maturity and fall individually.

man might have used its extremely hard grains for food. Shortly after
the publication of the Mangelsdorf-Reeves original paper, George
Beadle, who did not accept the hybrid origin of teosinte, showed
that grains of teosinte could be popped like those of popcorn and
pointed out that primitive man could have used it in such a fashion.
Early written accounts have since been found indicating that the
grains were parched for eating. That man ever took such a seemingly
unpromising food plant and made it into one of the world's greatest
cereals may strain the belief of some people, but it appears more and
more likely that teosinte is indeed the plant that gave rise to maize.

Although presumably maize was a most mutable plant, it must have required considerable conscious selection by man to have produced the great number of races that existed when it was discovered by the Europeans. The development of a wild grass into the "happy monster" that once fed most of the people of the Americas was a most significant achievement of primitive plant breeders. The development of hybrid corn in this century ranks as one of the most outstanding developments of the modern plant breeder.

The story of hybrid corn has been told many times and in many places but it bears retelling. First, some general remarks about hybridization are in order to set the stage. Hybridization has already been mentioned several times in this book—natural hybridization between species and within species and man-made, or artificial, hybridization for plant and animal improvement. In either natural or artificial hybridization, genes may pass from one species or variety into another by means of backcrossing the first generation (F_1) hybrid to one of its parents and the new types resulting may be exploited by man. Such gene transfer may occur from many hybrids between species but not from all of them by any means for some hybrids are sterile or nearly so, like the mule. On the other hand, hybrids within a species, for example between two different varieties, are generally fertile as is true of hybrids between different varieties of maize. One other important characteristic of hybrids has already been mentioned: First-generation hybrids frequently show a greater vigor than either of their parents. This hybrid vigor, also called heterosis, is seen in the mule and was to prove to be of the greatest significance in corn.

Two of our greatest biologists, Charles Darwin and Gregor Mendel, figure, at least indirectly, in the production of hybrid corn. Darwin in some of his researches found that continual inbreeding of plants reduced vigor whereas crossing different varieties increased vigor. Darwin even used corn in some of his experiments, although nothing more was to come of it at the time. Mendel, of course, supplied us with the laws of genetics, which, after their rediscovery in 1900, made it possible for man to use deliberate planning in developing better plants and animals.

The history of hybrid corn begins with a man who had been much influenced by Darwin, William James Beal, who in 1877 made the first controlled crosses in maize in an attempt to increase the yield. In his work at Michigan Agricultural College, now Michigan State University, he proved that yields could be increased by crossing different varieties. Shortly after the turn of the century, at Cold Spring Harbor on Long Island, New York, George Harrison Shull, who was following up some of Mendel's discoveries, and at the same time, Edmund Murray East at the Connecticut Experiment Station, began studying the effects of inbreeding. Their self-pollinated maize plants became weaker after each generation, but they found that, if two inbred varieties were crossed, great vigor could be restored in this single step. The final touches for the ultimate commercial production of hybrid corn were provided by D. F. Jones, a student of East at Harvard, who went to the Connecticut Experiment Station in 1914. The inbreds* were very weak plants and produced rather small ears with few seeds. Thus if inbred A were pollinated by inbred B, the seeds produced on A would be used for hybrids, and the small number of seeds didn't make it of very practical use. Jones devised a double cross using four inbreds, A, B, C, and D. A was crossed with B, and C with D. The hybrids AB and CD were then grown and a hybrid made between them. This plant, being a hybrid, produced a large number of seeds, and the double crossed seed, when grown, still produced plants showing extreme vigor and uniformity. By utilizing the double cross it was now possible to convert a few bushels of single crossed seed into a 1000 bushels of double crossed seed, which could then be released to the farmer, making it economically feasible to grow hybrid corn. Thus through the work of several scientists in addition to those mentioned here, and practical farmers as well, hybrid corn became a reality. The double cross method for the production of hybrid corn seed is still employed, but gradually as better inbreds were developed single crosses have come more and more in use for the production of hybrid seed for direct sale to the farmer. In 1970 seventy-five percent of the hybrid corn grown in the United States was derived from single crosses.

By careful selection of the inbreds, a breeder can produce high-yielding maize hybrids for many different climatic zones. Special characters, such as resistance to disease and tolerance to drought, can be incorporated as well as special morphological features. Plants were developed that produced two or three ears to the plant, instead of

---

*In fact, the inbreds were such sick-looking plants that in the early days of hybrid corn development in at least one agricultural experiment station they were grown in out-of-way places where farmers would be unlikely to see them. The breeders felt that if the farmers saw the inbreds they would think the breeders were working in the wrong direction.

one, uniformly placed on the stalk to make harvest by mechanical pickers practical. Plants with stiff stalks were produced to withstand mechanical picking.

Hybrid-corn seed must be purchased anew each year, for if a farmer saved seeds from one year's crop for planting the next, he would lose much of the vigor and uniformity found in the original hybrids. Thus the production of hybrid corn seed became big business in the United States. Today in terms of dollars it is larger than that of the steel industry, with four or five companies producing most of the hybrid seed grown in this country.

The changeover to use of hybrid corn was slow initially. Only one percent of the corn produced in the United States in 1935 was hybrid, but today virtually all the corn grown is hybrid. It soon made possible the production on three acres what used to be produced on four. Equally important is the fact that hybrid corn has made possible a drastic reduction in the number of man hours required to produce a bushel of corn. The increased yields with reduced manpower were particularly important during World War II when, under serious depletion of the labor force, the amount of corn produced decreased by only 10 percent. Since the war yields have continued to increase dramatically, more than doubling in the space of 20 years. It should, of course, be obvious that other scientific advances, improved fertilizer applications and more efficient mechanization, for example, have contributed to its great success.

Hybrid corn has become important in other parts of the world and is gradually replacing much of the old types grown in Latin America with great increases in yield. Yugoslavia became a producer by importing hybrid corn seed from Iowa. Seldom can a plant well adapted to one area be an immediate success in another, but Yugoslavia, being at the same latitude as Iowa and having a climate not too dissimilar, was able to take immediate advantage of the introduced seeds.

In the production of seed for hybrid corn the seed companies grow a row of one inbred, A, to be the male parent, next to another, B, that is to serve as the seed parent. In order to insure that self-pollination of B does not occur, it is necessary to remove its tassels before pollen is shed. In the early years of seed production, this operation of detasselling was done by hand, college students often being employed for the work. One of the recent advances in hybrid-seed production has been the elimination of detasselling through the discovery of plants that produce no pollen. The factor for pollen sterility, referred to as

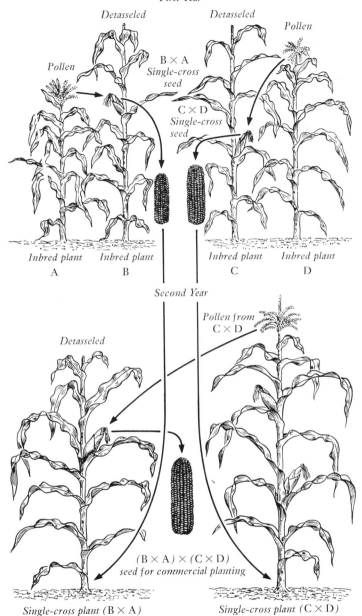

First Year

Detasseled

Detasseled

Pollen

Pollen

B × A
*Single-cross
seed*

C × D
*Single-cross
seed*

*Inbred plant
A*

*Inbred plant
B*

*Inbred plant
C*

*Inbred plant
D*

Second Year

*Pollen from
C × D*

Detasseled

(B × A) × (C × D)
seed for commercial planting

Single-cross plant (B × A)

Single-cross plant (C × D)

Figure 5-19
Double cross method of producing hybrid corn seed. The inbreds shown at the
top are crossed to produce seed to give the single cross hybrids shown below.
The single cross plants are then crossed to produce seed to be sold to the farmer.
(From "Hybrid Corn" by Paul C. Mangelsdorf. Copyright © 1951 by Scientific
American, Inc. All rights reserved.)

cytoplasmic male sterility, resides in the cytoplasm of the cell rather than in the nucleus and is inherited quite different from ordinary Mendelian genes. Since all plants derived from a male sterile parent produce no pollen, it is necessary for a breeder to introduce a gene into the hybrids to make the next generation pollen fertile. Virtually all the hybrid seed now sold in the United States is produced by utilization of plants carrying the male sterility factor along with genes for restoring fertility when grown by the farmer.*

It will, perhaps, come as no surprise to learn that nowhere in the world is more maize produced than in the United States. Argentina is second in production, and China is probably third. Maize is also of considerable importance in many other parts of the Old World, for in spite of the reluctance to accept it in some places it, next to tobacco, was the New World plant most widely adopted following the discovery of America. Although its cultivation in the United States is concentrated in a relatively few states, extending from western Ohio to eastern Kansas and Nebraska, maize is the single most important crop plant in the United States. It ranks behind several other crops in export value. So obviously most of it is used at home. In fact, some 80 percent of the crop never leaves the farm but is used directly for livestock food. The small amount that does go to market is extremely important and yields more products than any other cereal. Practically all parts of the corn plant are used in some way, but only the grain has significant usage. A few years ago a survey of a supermarket revealed that the grain was used in one way or another in the preparation of 197 different food products, and, if the drug counter had been included in the survey, many other products would have been added, for corn starch is used as a filler for aspirin and other tablets and corn syrup is used as a base for cough medicines. Some 200 million bushels go to the corn refineries, or "wetmillers," every year where the oil-bearing germ, or embryo, is removed, then the outer covering, the gluten, and finally the starch, which constitutes more than half of the kernel. The starch is used directly as starch or is converted into syrups or sugar. The starch is sold for use in food, drugs, cosmetics, or for other industrial uses. The syrups and sugars also find a great number of uses in foods, in the preparation of medicine, in chewing gum, and in making cold-rubber tires. The solubles from the steep

---

*Hybrids containing the male-sterile cytoplasmic factor were seriously affected by blight in 1970; see footnote page 198.

water left after removal of the other substances and the glutens find many uses, such as in animal feeds and in the manufacture of plastics. The refined oil from the germ is used in cooking, in salad dressings, and in making of margarine. Perhaps other plants, for example, soybeans and coconuts, have more uses than maize, but certainly maize would run them a close race.

Research is constantly underway to find new uses for maize and to improve the plant, for maize has a badly balanced protein and, like rice, a lower protein content than wheat. It is also a poor source of certain vitamins. The deficiency disease, pellagra, caused by a shortage of the vitamin niacin, is often found among people whose diet is inadequate. Large amounts of the amino acid tryptophan have a niacin-like affect. Since maize is relatively deficient in both the vitamin and the amino acid, pellagra is often common among people who depend on maize as their staple. In 1964 scientists at Purdue University discovered that a mutant, called opaque-2, which had been known for more than a quarter of a century, had significantly higher lysine and tryptophan content than other types. Increases in these two amino acids greatly improves the nutritive value of maize, and in a protein-hungry world this could be more significant than any new industrial use for maize. Field tests in Colombia have already indicated that the new maize may have a significant impact on protein deficiency in that country.

As a botanist, I cannot close this discussion without mention of another important, although sometimes overlooked, use of maize—in teaching and research. In the development of modern genetics maize has been to the botanist what the fruit fly *Drosophila* has been to the zoologist. It is a particularly valuable plant because its chromosomes are exceptionally large, which has allowed detailed studies that have contributed greatly to our understanding of inheritance.

Sugar Cane

Quite unlike the grasses treated previously, sugar cane, *Saccharum officinarum*, does not owe its importance to its seed and fruits. In fact, this plant, which is valued for its stem, often sets few or no seeds. In prehistoric time man found that by chewing the stem he could obtain a sweet juice, as he still does in many places. Domestication is thought to have occurred in New Guinea or Indonesia. The plant spread

Figure 5-20
Cutting sugar cane in Ceylon. (Courtesy of FAO.)

throughout much of southeastern Asia. Europe learned of it through the travels of Alexander the Great when one of his men reported that in Asia the people were able to obtain honey without bees! Sugar cane did not become well known to the western world before the fifteenth century, and since that time it has spread throughout much of the world's tropics, Columbus is credited with having introduced it into the New World. Until the advent of the sugar-beet industry in the nineteenth century, sugar cane was the only source of sugar for most of the world's inhabitants. In this century, sugar cane has had an important influence on the economy of many countries, of which Cuba has been most prominent in recent years.

Sugar cane is a tall plant, not uncommonly reaching heights of ten to fifteen feet. Since it is a perennial, it will produce new stems for many years, but replanting is often done after the third harvest. Propagation is done vegetatively by using pieces of the stem. Extraction of the juice is accomplished by means of rollers. This is followed by purification that removes many of the "impurities," some of which are of nutritive value. The juice is then concentrated by evaporation. A boiling follows that leads to a crystallization yielding raw brown sugar and molasses. The brown sugar is then refined to produce white sugar. Except for the refining, all the steps in the processing of sugar are usually done on the plantation where the sugar is grown. The refining is frequently carried out in the importing countries of Europe or in the United States.

The refined sugar is one of the purest products to reach our tables and is one of our best energy sources. It has been said, however, that sugar contains "empty calories," by which is meant that it supplies man with nothing but pure carbohydrate. A diet consisting mostly of sugar can lead to malnutrition. Thus we have the "sugar babies" of the West Indies who are fed mostly on sugar and suffer protein malnutrition as a result. Sugar cane is said to give the highest yield of calories per acre of any plant, and in spite of their being "empty," there are some who are predicting that we shall have to turn to sugar as our chief calorie source as the world's population expands.

6

Poor man's meat: the legumes

And let them give us pulse to eat, and water to drink.
Daniel 1:12.

Man's first cultivated plants in the Old World, as we have seen, were grasses—wheat and barley—but they were soon joined by plants that added valuable supplements to his diet—lentils, peas, and vetches, all members of the botanical family Leguminosae. From the archaeological record we know that man was collecting seeds from wild plants of these species at the time he was beginning to cultivate the grasses, and their cultivation began not long after the domestication of the cereals. Cultivated lentils appear in the archaeological record in deposits accumulated before 5000 BC. The evidence from the New World indicates that beans, another legume, were one of man's first cultivated plants in the Americas, with one species going back to 5000 BC or earlier in Mexico and another to about 3000 BC in Peru, both antedating the appearance of maize in the regions. In the Far East another legume, perhaps destined to become more important than any other, the soybean, became an early domesticated plant.

It can hardly be an accident that these plants were among man's first domesticates, for their seeds are an excellent food and among the highest of all plant foods in protein content. Even though the cereals are given the credit for making man's civilization possible, it must be pointed out that he could not have advanced nearly as rapidly without the legumes. Not only are the legumes high in protein, but their amino acids neatly complement those of the cereals; if man ingests legumes and cereals together, he comes far closer to obtaining "complete" protein than from eating any plant food alone. Thus wheat plus peas, maize plus beans, or rice plus soybeans come close to filling man's protein needs. But in a way, the domestication of complementary food plants must be partly a "happy accident," for primitive man knew nothing of proteins or amino acids, only that the seeds satisfied his hunger and didn't make him sick.

A passage from the King James' Bible has been used to prove an early appreciation of the food value of the pulses, a name used for the edible seeded legumes in England. After he was carried to Babylon, Daniel was ordered to eat the king's meat and wine. He resolved, however, not to defile himself by doing so and proposed a test. He and the servants would eat pulses and water while the other captive youths would be given the king's meat and wine and a comparison would be made at the end of ten days. At the end of the time, the Bible tells us, "their countenances appeared fairer and fatter than the children which did eat the portion of the king's meat." The argument for the value of the pulses, loses its force, however, in modern translations of the Bible in which "meat" is given as "rich food" and "pulses" as vegetables."

If we were to rank the families of plants in order of their importance to man, certainly the legumes would stand close to the grasses, for they serve man not only as food but also perhaps in a greater variety of other ways than do the grasses. One of the greatest services they perform is the fixation of atmospheric nitrogen. When plants and animals die and decay, their nitrogenous compounds are broken down by bacterial action. Some of the nitrogen released is made available to other plants in the form of nitrates, but much of it escapes to the atmosphere and is not directly available to plant life. Most members of the legume family, however, are able to obtain nitrogen from the atmosphere through the action of special bacteria that live in nodules on their roots. Very few other plants, no cultivated food plant among them, have this ability. Although man had

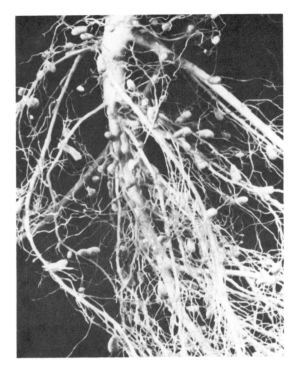

Figure 6-1
Legume root with nodules. (Courtesy of USDA.)

long recognized the value of legumes to the soil, not until late in the nineteenth century was the scientific explanation for their importance formulated. The bacteria, living in the nodules on legumes, belong to several species of the genus *Rhizobium*, and derive their energy from their host and in turn provide nitrogen to it in a symbiotic, or mutually beneficial, relationship. The nitrogen fixed goes into the production of protein, which is then available to man through the seeds, and to animals through leaves and stems as well as the seeds. It is also returned to the soil where it may be used by other plants. Although the nitrogen-fixing bacteria are widely distributed on wild and weedy legumes, farmers today, when planting a new field to legumes, often inoculate the seeds with a commercial preparation containing the bacteria to insure the presence of nodules or to increase their abundance. Before commercial fertilizer was widely available, legumes were frequently employed in crop rotation schemes to build up soil fertility and still are so used in many parts of the world.

The legume family is an extremely large and cosmopolitan one and all the major continents have provided members of it that have become foods for man. These plants are most commonly known by the names pulses, peas, or beans, but not all plants called beans belong to this family; castor bean, for example, is not a legume. The seeds are the part most commonly eaten, and, like the cereal grains, most of them can be fairly readily stored for future use. Like the grasses, many of the domesticated legumes have lost their natural means of seed dispersal. Most members of the family produce a seed pod, botanically known as a legume, containing a row of seeds. In wild species the pod is frequently dehiscent at maturity, splitting along both sides, often forcefully expelling the seeds to some distance from the parent plant, but many of the domesticated forms have pods that are indehiscent*—an advantage to man in that it makes it easy for him to collect all the seeds at the same time, but a disadvantage to the plant in that it interferes with the dispersal of seeds under natural conditions. The domesticated plants also have considerably larger seeds than their wild relatives, just as was found for the fruits of the cereals.

The food value of the seeds is high: they have about the same caloric value per unit weight as cereals, and are a fair source of some vitamins and minerals. However, as already mentioned, it is their protein content that is striking, generally ranging from 17–25 percent, about double that of most cereals, to a high of 38 percent in the soybean. The protein quality is not as good as that of meat and other animal products, but meat is not regularly available to much of the world's people. The legumes are thus often thought of as poor peoples' food, and many of the world's poor could use more of them. Consumption, as might be expected, is highest in India where both poverty and religious restriction on meat contribute to their great use, and it is also very high in Latin America, where beans are frequently served with all meals.

In addition to the seeds, the green, unripe pods are often consumed,

---

*As a young professor of botany, I learned this the hard way. I brought green beans to class to illustrate the legume type of fruit and I told my students that when they dried they would split open. They never did.

green beans being one of the most common vegetables in the United States. Less well known is the fact that the ripe pods of several of the tree legumes have high sugar content and are eaten by man or livestock. The pods of carob, or St. John's Bread, the "locusts" of John the Baptist in the Bible *(Ceratonia siliqua)*, native to Syria, are still eaten by man for their sweet taste, although more widely employed as livestock feed. The sugar obtained from the pods is used in preparation of chocolate substitutes including a candy bar that, according to the advertisements, tastes like fine Dutch chocolate. The genus *Inga* in Latin America has extremely large pods, some nearly a yard long, containing seeds covered with a sweet white pulp. The pulp is sucked or chewed off and the seeds discarded. This was one of the few sweets known to the people of this region until the introduction of sugar cane. The pods are sold in many markets in tropical America, and the trees are among the preferred species cultivated to shade coffee plants. Young sprouts or seedlings of various kinds of beans are a popular food in some places, particularly in the Far East. Several legumes have edible roots, one, *Pachyrrhizus*, the yam bean, or jicama is cultivated primarily for its root in parts of Latin America as well as in the Orient where it was introduced.

Legumes rival grasses as foods for animals. When we speak of hay or forage, we generally are referring to both grasses and legumes, and sometimes to other plants as well. Leaves of legumes are particularly valuable for animals for the same reason that the seeds are for man— their high protein content. Alfalfa, or lucerne *(Medicago sativa)*, is one of the best, since it is both high yielding and high in protein and is probably the oldest cultivated forage plant. Seeds more than 6000 years old have been found in archaeological deposits in Persia. Alfalfa spread to Europe in early Christian times and in the last 150 years has made its way to most other parts of the world. Although it had been introduced before the Revolutionary War, its importance as a crop in the United States dates from 1850, when seeds were brought to California from Chile by a gold miner. For a long time it was thought to be adapted only to the western states, but in this century varieties were found that would grow in other states and now the north central states rival California in total tonnage produced. Alfalfa has been recommended as a human food and is sold by some "health food" stores, but it has never caught on—its flavor leaves something to be desired. Cigarettes made of alfalfa leaves have had the same fate and for the same reason.

Several species of clover, lespedeza, kudzu, vetches, and some of the plants also grown for food, such as the field pea, are other important forage plants of this family. Most of these plants are also of value in erosion control, soil building, and as food for wild life, including nectar for bees. Although perhaps not equal to the grasses in controlling soil loss, they may serve as better soil builders because of their ability to fix nitrogen.

The family has also furnished us with some of our favorite ornamentals, such as sweet peas, lupines, and scarlet runner beans among the herbs, the vine wisteria, and a great many trees—some grown for their flowers, some for their graceful leaves, others for both. In the temperate zone of the United States the native red bud and honey locust, and the Asian *Albizzia*, are extensively cultivated. The tropics furnish a great many more, many with brilliant flowers, such as poinciana. Some mention must also be made here of *Mimosa*, the sensitive plant, native to tropical America, but widely grown in greenhouses for its sensitive leaves, which close to the touch. Countless students of botany have seen the plant used to illustrate plant movements.

A number of the trees, particularly tropical species, are valuable sources of wood, one of them being the much prized rosewood. Other members of the family produce gums and resins that are used in medicine and in varnishes. Dye plants are also known in the family; indigo, until replaced by synthetics, was particularly valuable. Flavorings such as licorice and tonka beans are still of some importance.

On the negative side, in addition to providing a number of weeds, the legume family also includes a number of poisonous plants, some of which are known to have caused deaths in both humans and domesticated animals. A few of these deserve our attention.

The jequirity bean or rosary pea, *Abrus precatorius*, a woody vine that is rather widespread in the tropics, has been newsworthy on several occasions in recent years, with warnings about the toxic effects of the seeds appearing. Beads made from the attractive bright red and black seeds have been brought home by tourists or imported for sale by stores. The seeds make beautiful necklaces, and are dangerous only if they are chewed and swallowed. The active principle, abrin, is one of the most toxic substances known, one seed containing enough to kill a person. It is reported, however, that in parts of Africa, the cooked seeds are eaten without harm!

A few of the important leguminous food plants are known to be somewhat toxic under certain circumstances or to certain persons.

122

Figure 6-2
Novelty pins with jequirity beans used for eyes.

The distinctive taste of the lima bean is due to a cyanogenetic gluco-side, and although beans grown and sold in the United States have only very small amounts, some varieties from the West Indies have enough to be considered dangerous.

Two diseases in humans, lathyrism and favism, are associated with legumes. The former results from consuming large amounts of the grass, or Indian, pea, *Lathyrus sativus*. Eating the grass pea as part of an ordinary diet causes no damage and the plant is in fairly wide use as food in both Asia and Europe. Difficulty usually arises when other food supplies are scarce, and people are forced to subsist almost en-tirely on the legume. As the grass pea does well on poor soil and withstands drought, it may be in plentiful supply when other foods are lacking. Excessive consumption of it leads to a paralysis of the lower limbs that may be permanent. The disease has been most severe in India, but it was known in ancient times from Europe. Favism, an acute anemic condition, results from eating uncooked or only par-tially cooked broad beans, *Vicia faba*, also known as horse, or English, bean, or from inhaling pollen from the plant. The disease affects only males of Mediterranean origin, and it is now thought that an inherited biochemical abnormality is responsible. Neither the grass pea nor the broad bean is eaten to any great extent in the United States.

The genus *Lupinus*, in addition to furnishing us ornamentals, has given us species used for food. Several species were domesticated in the Old World and one in the New for their seeds. Lupines contain

Figure 6-3
Cultivated lupines *(Lupinus mutabilis)* in highland Ecuador.

alkaloids that are toxic. In the Old World man has selected alkaloid-free varieties for consumption by himself and his animals. One of these may be purchased at Italian markets in the United States under the name lupini bean. The Andean food species, called chocho or tarwi, *Lupinus mutabilis*, however, does contain large amounts of the bitter alkaloid. To make the seeds palatable as well as safe to eat they are soaked in water for several days, which leaches out the alkaloids. The chocho was once an important protein source in the high Andes in the area where the starchy potato was the major food source. Although it is still fairly commonly cultivated in parts of the Andes, it has been replaced in many areas by the broad bean, which also does well at high elevations and does not need the lengthy preparation prior to eating.

Many of the wild plants in the family are also poisonous. Prominent among them are the locoweeds, various species of the genera *Astragalus* and *Oxytropis*, which are widespread in the western United States and which have been known to cause death in livestock that have grazed upon them.

Although the family has supplied plants poisonous to man, it has provided him with others that contain substances lethal to insects but relatively harmless to humans. One of our safest insecticides, rotenone, comes from species of the South American genus *Lonchocarpus* and its Asiatic counterpart, *Derris*. Long before their insecticidal properties were discovered they were used, as were a number of other plants, by primitive man as fish poisons. The plants were pounded and placed in dammed-off waterways. The substance released from the plants stupified the fish and as they came to the surface they were gathered; the poisoned fish could be eaten without harm.

The list given here by no means exhausts all of man's uses of members of this family, but it should serve to illustrate this plant family's broad significance. We may now return to some of the food plants and examine them in greater detail. Three of them, the common bean, soybean, and peanut, rank high among man's most important domesticated species and several others continue to occupy prominent roles as food plants.

As has already been pointed out peas and lentils are among the very early cultivated plants of the Near East. They are still important there and throughout much of the world. The cultivated pea *(Pisum sativum)* comprises two main races or varieties—the field pea, now used mostly for forage and for dried peas, and the garden pea with its high sugar content, considered by some to be the aristocratic food plant of this family. Fresh peas were not popular before the seventeenth century at which time they became esteemed in the court of Louis XIV. Whole pea pods used like green beans are a favorite food in the Far East, and are becoming increasingly popular in the United States. Biologists, of course, need no reminder that it was the pea that served as Mendel's research material for working out the laws of genetics.

The broad bean, previously discussed, chickpeas or garbanzos, *Cicer arientinum*, and the cowpea or black-eyed bean, *Vigna sinensis* are other fairly important domesticates of Old World origin. The cowpea is widely used in the deep southern part of the United States. Another species, *Vigna sesquipedalis*, called asparagus or yard-long bean, is frequently offered by seed companies as a novelty, but those who grow it can hardly expect the beans to reach a yard in length. The specific epithet, translated as a foot and a half, comes closer to the truth.

No genus in the family has provided more edible species than has *Phaseolus*, and this is the group to which we generally refer when we speak of beans. Several of the species are native to Asia and were brought into cultivation there, while others are American in origin. The domesticated species of the two hemispheres are not particularly closely related; in fact, recent taxonomic studies indicate that the Old World species should be placed in the genus *Vigna*. Man had nothing to do with their transfer from the one area to the other until after Columbus' discovery of America. The genus, like many others, attained a nearly worldwide distribution previous to man's interest in them as food plants.

The Old World members of *Phaseolus (Vigna)*, known as black gram or urd bean, the golden gram or mung, the adzuki bean, the rice bean, and the moth or mat bean, are used mainly in India, China, Japan, and some of the neighboring areas. Their seeds are smaller than those of the American beans, and this may be one reason they have not been more widely adopted throughout the world. The germinated seeds of the mung bean are the widely used bean sprouts of Chinese cookery.

Of the four American domesticates of *Phaseolus*, two are of limited use today. The scarlet runner bean, *Phaseolus coccineus*, originally domesticated in Mexico, is grown in the United States mainly as an ornamental but the large beans are a good food. The tepary bean, *Phaseolus acutifolius*, which was very important in prehistoric times in the American southwest, although more drought resistant and less gas-forming than the other beans, is not grown commercially.

The lima, butter, or sieva bean, *Phaseolus lunatus*, is fairly extensively cultivated today. This species was first domesticated in South America and a large-seeded form is known archaeologically from Peru dated at about 3000 BC. The earliest remains from Mexico are much later, about AD 800, and are small-seeded forms. The plant that may be the ancestral type, while not common, is fairly widely distributed in tropical America today, and it is thought that the large- and small-seeded types of lima beans may represent independent domestications. Although the lima beans in markets in the United States are nearly always white when mature, a great variety of colors are known in those from Latin America. Speckled lima beans were used to convey messages in ancient Peru.

The most widespread and widely used bean today is *Phaseolus vulgaris*, known by a great variety of names, common bean, navy

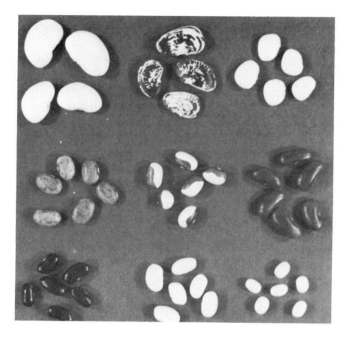

Figure 6-4A
Some bean varieties sold in the United States. Top row: lima
beans *(Phaseolus lunatus)*; Large Lima, Calico, Baby Lima.
Bottom two rows: common beans *(Phaseolus vulgaris)*; middle
row: Pinto, Yelloweye, Red Kidney; bottom row:
Black Valentine, Great Northern, Small White.

bean, and kidney bean being some of the most commonly used in
English. Our snap, or green, beans and wax beans are simply varieties
of this species whose pod is eaten in the immature stage. This bean
was domesticated in Mexico prior to 5000 BC, and its wild ancestral
type, which still grows there, has only recently been identified. The
domesticated bean reached Peru before the beginning of the Christian
era.* It also spread northward along with corn and squash, and these
three Mexican plants became the principal agricultural plants of
North America in prehistoric times.

Today numerous varieties of the common bean are known; one

---

*See footnote on page 10.

Figure 6-4B
Harvesting green beans *(Phaseolus vulgaris)* in New York. (Courtesy of USDA.)

estimate places the number at 500, only a few of which are grown in the United States. This bean is widely used in various parts of the world, and it still retains great importance in Latin America, where in different areas people have their own favorite varieties—red beans, or black beans, or other types. Mashed beans for breakfast are not un-common in parts of Latin America and not infrequently beans, as previously mentioned, along with rice, are served with every meal. Because the plants are vinelike, the Indians commonly planted beans among their maize plants, so that they could use the stalks to climb. Many beans are still grown this way but poles have replaced the maize for supports in many areas. A mutant type with a dwarf or bush habit that needs no support has become a favorite in vegetable gardens in the United States.

The two other legumes not yet discussed, the peanut and the soy-bean, are in many respects more significant than any of the foregoing, not only because of their exceptionally high protein content, but also for their high content of oils, which has given them numerous uses in industry as well as in food. The peanut, *Arachis hypogaea*, also called groundnut, groundpea, goober, pender, and many other names, is a rather unusual plant in that its fruits are produced under ground. The stalk of the flower elongates after fertilization and pushes the developing pod under the soil. The pod, or the shell, is the fruit and the peanuts are the seeds.* For a long time it was thought that the peanut was indigenous to either China or Africa but its true homeland is now clearly established to be South America. From there it was carried to other parts of the world in post-Columbian times. Its entry into the United States, however, was by way of Africa, being brought to Virginia by slaves. Archaeologically, the peanut is known from coastal Peru in the second millenium BC, but recent botanical studies have indicated that the place of domestication was most likely the foothills of the Bolivian Andes. How early it was domesticated there and how it reached Peru pose interesting questions for the archaeologist and botanist.

Peanuts are reported in archaeological deposits in Mexico at Tehuchan in levels dated at about the beginning of the Christian era, but they never became as significant in Mexico as they did in parts of South America. The Mexican Indian name, *tlalcacahuatl*, means "ground cacao," presumably from the fruit being borne underground and its resemblance to the fruit of cacao, the chocolate plant. From this we get the modern Mexican name, *cacahuate*, although the word *mani* which the Spanish picked up for the plant in the West Indies, is more widely used in other parts of Latin America.

Today the peanut like its relative the soybean and many other crops—coffee, rubber, and quinine, for example—is grown more extensively in other parts of the world than in its place of origin. China and Japan lead the world in production of peanuts, followed by certain African countries.† An attempt to grow the peanut on a huge scale in Africa led to one of the greatest agricultural fiascos in modern times. In 1946 the British government proposed to plant over 3 million acres of peanuts in East Africa for oil production. But clearing the land, the irregular rainfall, and disease proved to be serious obstacles and by 1951 only 65,000 acres were under cultivation. Later

---

*Although they may be sold with the nuts at the grocery store, the peanut is not a nut, botanically speaking.

†A native African legume that also bears its fruit underground, the bambarra groundnut, was largely replaced by the peanut.

Figure 6-5
Peanuts on freshly dug plant. (Courtesy of USDA.)

the project was abandoned but only after over 100 million dollars had been spent. As John Purseglove has pointed out, this should be a lesson to those who would try to make abrupt change in agricultural practice without sufficient research.

The peanut is put to far more uses in the United States than in any other part of the world, although this country ranks only fifth in world production. The plant is adapted to warm climates with light soil and is grown mainly in the southern states, with Alabama and Virginia responsible for half of the production in the United States. It spread in years when another crop was needed to replace cotton, which had been wiped out by the boll weevil. Only since 1850 has it become an important crop. In the last century most of the peanuts were used for roasting in the shell, and their prominence at circuses and baseball games can still be remembered. In fact, they were so much in demand that when one owner of a baseball park threatened to eliminate peanuts because of the work involved in cleaning up the shells, he almost had a rebellion on his hands. In the 1900's the mechanization of all operations connected with the growing and harvesting of peanuts was a strong impetus to increased production and since that time peanuts have been a major crop in the United States. Of the numerous varieties of peanut, only two, the Spanish and Virginia, are commonly known in the United States.

The principal use for peanuts is for human food, and a protein content of 26 percent makes it an extremely valuable one. In the United States the greatest amount goes into peanut butter, and countless children, of course, have subsisted almost entirely on peanut butter sandwiches, often by choice. Although Indians of South America had long ago made a similar product, its use in the United States dates from 1890 when a physician in St. Louis came up with it in a search for a nutritious, easily digested food for invalids, and from 1893 when Dr. John Harvey Kellogg, the health-food faddist of breakfast-cereal fame, made peanut butter so that some of his patients with poor teeth could take advantage of what he called "the noble nut." Peanut butter is very simply made, and the best grades are made simply by grinding the roasted and blanched nuts and adding 1–2 percent salt. Stabilizers are sometimes used so that the oil does not separate out. Considerable amounts of peanuts, some 70,000 tons annually, are also used in the manufacture of candy. Peanuts yield an oil that is esteemed as a cooking and salad oil and for the manufacture of margarine. There is considerable demand for peanuts in Europe and in many years the United States produces a surplus that is used for export.

A problem of considerable concern in recent years has resulted from the recognition that a mold infestation of peanuts produces aflatoxin, a harmful compound, that may be carcinogenic. Fatalities are known to have occurred in animals from eating infected nuts. Since the aflatoxin can also be found in processed peanuts and other agricultural products, humans are also subject to poisoning. Fortunately it is now recognized that by using proper harvesting and storage procedures, the fungus responsible for the aflatoxin can be eliminated.

George Washington Carver was instrumental in developing new uses for the peanut early in this century, realizing that the welfare of the·South could be greatly influenced by the plant. Today peanuts are important industrially and are used in the manufacture of a number of products from shaving creams to plastics. The residue from the oil-extraction process can be used for fertilizers and livestock food and could, with proper treatment, be used for human food. Peanuts are an excellent food for fattening hogs, and sometimes the animals are turned loose in the fields to root out the nuts for themselves. The shells can be used to make insulating filler and wallboard, as well as in other ways.

Although the peanut has come to be one of the world's important food plants, its rise to prominence hardly equals that of soybeans. The soybean, *Glycine max*, can hardly be considered a new crop. Although it didn't reach Europe before the eighteenth century and the United States until a century later, it was a very ancient and important cultivated plant in the Orient. Thus far there are no archaeological records to help us to establish when it was first cultivated, but its mention in Chinese literature before 1000 BC gives it a venerable age. The soybean is one of the richest foods known, containing 38 percent protein and 18 percent fats and oils. Unlike most members of the family, the beans are seldom eaten directly but are used, it is said, in 400 different ways in the Orient for food. A paste, curd, and "milk" made from the beans are used in great variety and the bean sprouts are used as a vegetable. Sauces of many different sorts are prepared by fermentation of the beans with addition of other ingredients, such as salt, wheat, and sometimes even decomposed chicken or fish, which further enriches the protein content of the food. A soft drink, Vitasoy, is now being made from it in Hong Kong. Today China is a distant second to the United States in world production of soybeans.

The widespread cultivation and utilization of soybeans in the United States in the space of a few decades has to be one of the most spectacular success stories in the recent history of agriculture. Soybeans were first grown in the United States in Pennsylvania in 1804, and although at the time it was mentioned that they should be more extensively cultivated, very little was done until the 1920's. Five mil-

A

B

Figure 6-6
A. Field of soybeans in the United States. (Courtesy of National Soybean Processors Association.)
B. Soybeans ready for harvesting. (Courtesy of National Soybean Processors Association.)

lion bushels were produced in the United States in 1925. Production reached 13 million bushels in 1933, 90 million in 1939, 299 million in 1950, 699 million in 1963 and increased to 1080 million in 1968. Little wonder that it has been called the "Cinderella crop." A number of factors contributed to the crop's rapid growth. The introduction of more than 1000 strains from the Orient in the late twenties provided material for the selection of improved varieties for different parts of the country. It was found that the soybean was well adapted to the corn belt, where it now ranks as second only to maize. The price and income support for other crops in the thirties encouraged the spread of the soybean. The development of specialized equipment for planting and harvesting was a factor in increased acreage being devoted to it, and the discovery that soybean meal was an excellent food for livestock, particularly poultry and hogs, made a large market readily available. At the same time new food and industrial uses for the soybean came into being. Today the United States produces more than 70 percent of the world's soybeans, about two-fifths of which is exported in one form or another, which makes it the United States' most important agricultural commodity in world trade.

Although it still has its greatest food use in the Orient, the plant has become increasingly used as food in other parts of the world, particularly as an additive for protein enrichment and as a meat substitute. By spinning the protein into long slender fibers and adding appropriate coloring, flavoring, and other nutrients, soybeans can be made into synthetic beef, chicken, bacon, or ham. The taste, except possibly for the synthetic bacon, is not yet very close to the real thing but improvements may be expected. The soybean is the world's greatest source of vegetable oil and the oil is of high quality. Most of it is used in the production of margarine, and some for shortening, salad and cooking oil. Soybean lecithin, a substance yielded in processing the oil, is used in many ways—to preserve flavors of other foods, to help disperse nonsoluble compounds in food, and in whipped toppings, cake mixes, and instant beverages. Eighty-five percent of the soybean crop in the United States goes into the production of food for man and livestock, the remainder finds wide usage in industry, perhaps greater than that of any other plant. It is used to make a glue that is the most widely used adhesive in binding plywood, and in the manufacture of enamels, linoleum, printing ink, and soaps—to name only a few.

Certainly many of the hungry nations could use this highly nutritious food but at present they get little or none of it. Although the soybean has been introduced into many tropical countries in this century, as yet it has met with little success in most of them. In some tropical areas of Africa where it is now grown, the beans are used mostly for export so it helps little in alleviating the local protein deficiencies.

7

The starchy staples

But don't forget the potatoes.
JOHN TYLER PETTEE, *Prayer and Potatoes.*

In an earlier chapter the statement was made that the cereals are the basic food of man. Many peoples, however, in former times and even today, living in areas poorly adapted to growing cereals have had to adopt some other plant as the mainstay of their diet. Several of these—chiefly, the potato, the sweet potato, the yam, manioc, and the banana—are still extremely important.

Although not closely related, all of them belonging to different botanical families, these plants have much in common. All are tropical in origin, although the potato comes from the highlands whereas the others are lowland plants. They are all propagated vegetatively, rather than by seed. The archaeological record for most of them is scanty or nonexistent, as is true for most tropical plants, which grow in areas where conditions for preservation of plant remains are poor. All of them with the exception of the banana produce their edible parts underground. They provide extremely high yields of carbohydrates and thus they supply good energy sources and a full belly, but all are sadly deficient in protein. A diet made up almost exclusively of any of them can lead to serious problems of malnutrition.

The Potato

The white or Irish potato, *Solanum tuberosum,* which rivals wheat in volume produced and value, had a long way to go before it became an acceptable food plant in Europe, but few plants have figured more prominently in Western history than has the potato. The story begins in South America. Wild potatoes are fairly widespread in the Americas, particularly in the Andes. They were probably found to be a valuable food when man first entered this area, and at some undetermined time, probably more than 4000 years ago, they came to be intentionally cultivated. With selection by man there was an increase in size, and the potato became the most important food plant in the high Andes, for it thrives at an elevation where few other cultivated plants will grow. Maize will not grow at elevations much higher than 11,000 feet and is not a particularly productive plant at that altitude, but potatoes do well at 15,000 feet. Frost, common in parts of the Andes, is not conducive to keeping potatoes, but the Indians found how to make it an ally. Potatoes were allowed to freeze at night and the next day the Indians would stamp on the thawing potatoes. This process, repeated for several days, removes the water, resulting in a desiccated potato, called *chuño,* which may be kept almost indefinitely and used as required. Thus was born one of man's original "freeze-dried" foods, in a sense the forerunner of instant mashed potatoes but with a somewhat different taste. Although the flavor of chuño is not pleasing to all foreign visitors in the Andes, it is still the staff of life to many Indians in highland Peru and Bolivia. The potato was cultivated throughout the length of the Andes in prehistoric time, but it did not make its way to Central America until introduced by the Spanish.

The first European on record as seeing the potato in America thought that it was a strange food and compared it to the truffle. It was introduced into Europe in 1570 but its acceptance was far from immediate. The Jerusalem artichoke, introduced from North America at about the same time as the potato, was welcomed and was soon regarded as food fit for a queen; today it is seldom used for food but the potato prevails. The slow acceptance of the potato was due to several factors. The newly introduced plant probably was not very productive at first, and the fact that it was recognized as a member of the nightshade family may have contributed to a reluctance to accept it. At this time the nightshade family was known in Europe mainly through its poisonous members—mandrake, henbane, and belladonna—

Figure 7-1
Irish or white potato plant. (From "The Late Blight of Potatoes" by
John E. Niederhauser and William C. Cobb. Copyright © 1959 by
Scientific American, Inc. All rights reserved.)

and the family had not yet provided any important foods. Even after the potato gained some acceptance as food, there were many who disapproved of it. One minister preached against it, stating that if God had intended it as food for man it would have been mentioned in the Bible. Several writers of the time condemned it for its flatulent or "windie" property, among other reasons. At the start of the sixteenth century it was thought to be an aphrodisiac, stemming from confusion with the sweet potato, but this, of course, may have increased its use among some people. Promoted by royalty in some countries, the potato gradually became more widely grown. Various wars that destroyed standing crops of other food plants at the time helped to increase its popularity, for the potato tubers, safe underground, could not be destroyed by burning as easily as a field of grain. The potato did not become really common in Europe before the eighteenth century, and there can be little doubt that it contributed to the increase of the population at that time, not because of any aphrodisiac property but because of its food value. World War I has even been blamed on the potato, since it was responsible for the increased population that has been considered a contributing factor to the war. Certainly it was the potato that helped to keep Germany alive during two world wars.

In the early part of the nineteenth century the potato had become the dominant food in Ireland. In fact, it was almost the sole food of the peasantry, and the average per-person consumption of potatoes was ten to twelve pounds a day. When a blight struck in 1845 and 1846, wiping out nearly the entire crop, famine followed. An estimated one and a half million people died as a result, and another million emigrated, many to the Americas. The blight disease was not understood at the time and only later was it recognized that it was caused by a fungus. Since that time plant pathologists and breeders have devoted a great deal of effort to controlling the disease and developing resistant varieties. Much has been accomplished but blight still causes serious losses in many parts of the world. It had been thought that eating blighted potatoes had no ill effect on humans, but recently it has been shown that a high incidence of certain kinds of birth defects characterizes areas where potato blight is common. In an experiment marmosets who ate blighted potatoes as part of their ration produced some offspring showing birth defects; a control group of the animals, fed an ordinary diet, bore all healthy young.

Although the potato is often referred to as a root crop, actually the part eaten is an underground stem, specialized for food storage and known botanically as a tuber. The eyes of the potatoes are buds and the so-called seed potatoes, used for propagation, are not seeds at all but portions of tubers containing an eye. Potato plants rarely flower in some northern areas, but they are capable of flowering and do so throughout much of their range. The attractive white, blue, or pink flowers produce small green berries, something like a small unripe tomato, that contain seeds. Although these true seeds are not used by the farmer, they are important to the plant breeders in creating new varieties. Vegetative reproduction, which is asexual, gives rise with rare exceptions to types exactly like the parent. Sexual reproduction by seed, on the other hand, allows for the production of new combinations, some of which may be superior to the parents.

The potato tuber, like most vegetables, is mostly water and contains 17–34 percent carbohydrate, small amounts of protein, a trace of fat and some vitamin C. Varieties commonly cultivated in the north temperate zone contain from 1–3 percent protein, but some in the Andes are known with 6 or 7 percent. Much of the greatest food value of the potato, which lies next to the skin, is removed, and therefore lost, through the practice of deep peeling. The potato reveals its relationship to the other nightshades not only in its flower and fruit but also in that it may contain the toxic alkaloid solanine. Potato tubers grown exposed to light turn green and produce solanine. Eating of the leaves or other green parts of the plants has caused poisoning in livestock and humans.

The English name potato for this plant is a mistake that goes back to the time of its introduction into England. The word potato is actually derived from *batata*, an Indian name for the sweet potato. There was considerable confusion over the new root crops being introduced into Europe in the sixteenth century, and the name of the sweet potato became attached to the white potato where it remains.

Today some 300 million tons of potatoes are produced annually. More than 90 percent of the production is in Europe with the Soviet Union, Poland, and Germany being the most productive. In addition to their use as food for humans, large amounts of potatoes are fed to livestock in Europe, and some go into the production of starch, used chiefly for sizing cloth and paper. Potatoes also serve as a source of alcohol, both for drinking and for industrial purposes. The potato reached the United States from Europe by way of Bermuda in 1621 rather than directly from South America. In the United States today

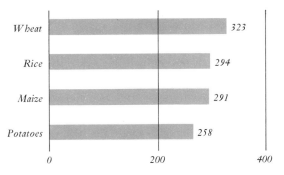

Approximate world production (million metric tons)

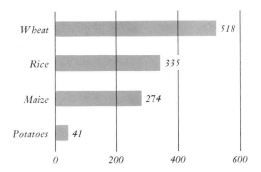

Approximate area planted (million acres)

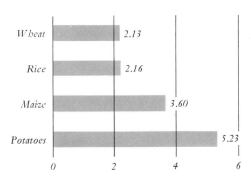

Approximate nutritional yield (million calories per acre)

Figure 7-2
The potato yields almost as much as any of the major cereals (top) but requires
much less acreage (center). Therefore, its nutritional yield per acre exceeds
that of these other staples (bottom). (Top and center graphs based on data from
USDA Foreign Agricultural Service for 1970–71 and do not include yield and
acreage from some of the smaller countries.)

about eight million tons are grown for direct consumption, and five million for use in other ways, much of it also reaching the table, as potato chips, frozen french fries, and dehydrated potatoes. A small amount is used for starch; the bulk of starch produced in the United States, however, is made from maize. The potato is, of course, still widely grown in the Andes and in Mexico and Central America as well. It has been grown in India since the seventeenth century, and later reached Africa where it is now grown in the highlands, particularly in Kenya. It cannot be grown successfully in the lowland tropics, where its place in man's diet is taken over by some of the other starch crops, particularly manioc.

Along with the studies aimed at improving the potato, considerable study is also being given to its origin. The potato is a tetraploid species and the identity of the wild species that are involved in its origin is still unknown. Many closely related wild species occur in the Andes. The suggestion has been made that it was first domesticated in the area of Lake Titicaca on the borders of Peru and Bolivia.

In the spring of 1970 five million pounds of potatoes were destroyed by fire in eastern Idaho. It was no accident, for the fire had been deliberately set by farmers in an attempt to raise the price. The same news report also carried the information that an estimated fifteen million Americans are underfed.

Sweet Potato

At the time of the Discovery of the New World, the sweet potato, *Ipomoea batatas*, was widely cultivated in tropical America and was also being grown on some of the Pacific Islands. In fact, the sweet potato had become the principal food of the Maoris in New Zealand. The presence of the plant on either side of the Pacific at such an early date poses several interesting questions—among them, how and when did it get across the ocean? The sweet potato is propagated vegetatively from stem cuttings or root sprouts. The plant under proper conditions may flower and set seeds although, as with the white potato, seeds today are used only in breeding. It has been pointed out, however, that seeds, either carried by birds or on floating logs, may have reached Pacific Islands and given rise to plants that man later discovered and cultivated. The suggestion has also been made that the sweet potato might have been independently domesticated both in the Americas and the Pacific area from similar wild species. While the introduction of seeds by some natural means or an independent domestication remains a possibility, it seems far more likely that man was responsible for the introduction of the sweet potato from the Americas to the Pacific region. There are two ways in which this might have occurred.

Ancient Polynesian voyagers, who are known to have been efficient seamen, could have traveled to South America, picked up sweet potatoes, then returned home with them and cultivated them. Or, and perhaps more likely, Peruvians, who also are known to have had seaworthy vessels, may have carried the sweet potato to one of the Polynesian islands, not necessarily intentionally but perhaps when driven off their normal course by winds. The prevailing currents are in the right direction for such a trip, and there is no need to account for a return voyage as with the Polynesians. There may be some question about whether the sweet potatoes would not have been eaten on the journey and, if not, whether they still would have been viable after such a long voyage, for sweet potatoes are rather perishable.

The presence of the sweet potato in both the Pacific region and South America does not necessarily mean that voyages across the Pacific were frequent in prehistoric times. If they had been, we would expect other plants to have been exchanged among the peoples of the two areas. Many such claims have been made for the spread of other plants, but except for two of them the evidence is unconvincing. The only other plants that did become established on both sides of the Pacific in prehistoric times are the coconut and the bottle gourd, both of which have fruits that float and hence are well adapted to dispersal by ocean currents. The voyage of the *Kon Tiki* by Thor Heyerdahl was an attempt to prove that primitive man could have crossed the Pacific successfully, and Heyerdahl and others have claimed that there were significant contacts in prehistoric times with an exchange of cultivated plants. There may have been such voyages, but, so far as plants are concerned, the only support for the idea of exchange appears to be provided by the sweet potato, and, of course, a single accidental voyage could have accounted for it.

Although the origin of the sweet potato has recently been given considerable attention, much of its early history in the Americas remains obscure. The sweet potato, like bread wheat, is a hexaploid and a Japanese botanist, Ichizo Nishiyama, a few years ago reported finding a wild hexaploid species, *Ipomoea trifida*, in Mexico that he

Figure 7-3
Sweet potatoes on vine.

believes is the ancestor of the domesticated plant. More recently he has identified the diploid and tetraploid species apparently involved in its origin. If he is correct, the sweet potato had its origin in Mexico. On the other hand, an anthropologist, Patricia J. O'Brien, from linguistic and anthropologic considerations, has recently concluded that the sweet potato originated in northwestern South America. In this connection, it is of interest that the sweet potato is known archaeologically from Peru but has yet to be found in other areas. The plant was also being grown in the Caribbean area when the Spanish arrived. The exact place or places of origin as well as its diffusion into other areas remains to be clarified.

The post-Columbian history, as is to be expected, is much better understood. Columbus took the sweet potato to Spain and it later became a common staple on ships on return voyages to that country. The Spanish also introduced it from Mexico to Guam so that it might be available as ship supplies on voyages to the far Pacific. As the plant is adapted to warm climates, it was not added to the cultivated plants in most of Europe. Sweet potatoes grown in Spain were imported into England where, for reasons not entirely clear, they were regarded as an aphrodisiac. Man has always wanted to attach aphrodisiacal properties to plants and it is only natural that such properties would become associated with exotic plants instead of those already well known.

The plant had various names in the Americas, *apichu* among the Quechua of Peru, *camote* in Mexico, and *aje* for starchy types and *batata* for sweet types in the Caribbean area.* Of these the last became the most generally accepted and was transferred to the white potato as was related above, which means that we now have to refer to the plant under discussion as the sweet potato. This name, of course, is not inappropriate since 3–6 percent of the carbohydrate is in the form of sugar, and this may increase in storage and with cooking. The plant provides 50 percent more calories than the white potato but generally has less protein (1.5–2.0 percent). It is a good source of vitamin A and minerals. In contrast to the white potato, the part eaten is a true root.

Although the plant is still widely grown in the Americas and has become an important crop in the southern United States, it is now more extensively cultivated in Africa and southeastern Asia. Japan now produces ten times as much as the United States, and it is the second most important crop in that country, rice, of course, being first. In addition to its use for human and livestock food, considerable amounts are used for alcohol production in Japan. In both Japan and Taiwan, the crop is regarded as "typhoon insurance," for when rice and other standing crops are destroyed the sweet potatoes will still be available for food.

The sweet potato is a member of the morning glory family. The flowers, seldom if ever produced under temperate-zone conditions, bear a close resemblance to the ornamental forms of the morning glory. Although more common at lower elevations, the plant can be successfully cultivated as high as 9000 feet in the tropics. The roots can't stand waterlogging and for this reason the plants are often

---

*The name *kumar*, or variants of it, was used in both Peru and Polynesia for the sweet potato. Until recently it was claimed that this was further evidence that the sweet potato was introduced by man from Peru to the Pacific area, for it would be most unlikely for the same name to have been independently chosen in the two areas. The Peruvian origin of the name has been questioned in recent years, and it has been postulated that *kumar* was actually of Polynesian origin and was introduced into Peru in the early post-Columbian period.

grown on ridges or mounds to provide good drainage. The vines rapidly cover the ground and hence the plants require little cultivation. In spite of the fact that for agricultural production the crop requires good drainage, the roots will sprout readily if placed in a jar of water and the vines will grow luxuriantly, and not infrequently such plants are grown as ornamentals in the home. In some areas the leaves as well as the roots have served as food.

Manioc

Manioc, cassava, and yuca* are some of the common names of *Manihot esculenta*, which is little known to most people in the temperate zones except in the form of tapioca, although it is one of the extremely important food plants of the tropics of both hemispheres. Although it is New World in origin, details about where and when it was first domesticated remain vague. Related wild species occur in both South and Middle America. Prehistoric remains are known from coastal Peru, but it is generally agreed that the plant was introduced there, perhaps from Venezuela or Brazil. The plant may have been independently domesticated in Middle America as well as in South America, but an introduction from South America to Middle America seems more likely. Manioc belongs to the Euphorbiaceae or spurge family that, among many other plants, includes the Pará rubber tree, our best source of natural rubber, and the poinsettia, a well-known Christmas ornamental. The manioc plant is rather tall, at times reaching fifteen feet, with divided leaves. The edible part is the tuberous root, which somewhat resembles a sweet potato, but is usually much larger—some grow to be a yard long and to weigh several pounds. Numerous rather ill-defined varieties exist that are generally divided into two groups, the sweet maniocs and the bitter. The latter contain higher concentrations of poisonous cyanogenetic glucosides than the former and require special preparation through grating, pressure, and heat to make them safe to eat. One wonders, of course, how man originally discovered that this plant, as well as several others that are toxic until specially prepared, could be made edible.

Manioc was taken to Africa from Brazil by the Portuguese in the sixteenth century but it did not spread widely there until the twen-

---

*Not to be confused with *Yucca*, an entirely different plant.

A

B

Figure 7-4
A. Manioc plantation, Ecuador. (Courtesy of FAO.) *B.* Manioc roots (front left) for sale in Quito market. Also shown are fresh maize for roasting on the cob, limes, and small hot *Capsicum* peppers (front center).

tieth century when its cultivation was encouraged. It was found that manioc was not damaged by locusts, a serious pest of crops in many parts of Africa, and that the bitter varieties could be grown in areas where wild animals would destroy other crops. The plant also grows better on poor soil than any other major food plant. As a result Africa today produces as much as, or more than the rest of the world combined, which in one sense is unfortunate, since the roots contain little protein and their wide use has contributed to malnutrition. The plant reached the eastern tropics somewhat later than Africa, and presently is most important in Indonesia, where it ranks as the third most important crop, after rice and maize. Some there is grown for export, although most of it is used locally for food, as is true in most other areas where the plant is grown. World production is estimated at more than 80 million tons.

Manioc is a lowland tropical crop, although sometimes it is grown at elevations as high as 6000 feet. It will grow in somewhat arid regions as well as in regions with fairly high rainfall. Stem cuttings are used for propagation, and are simply stuck in the ground at an angle in fields prepared by slashing and burning. The plants are then more-or-less left to themselves. Some varieties mature their roots in as little as six or seven months, and in others roots may continue to increase in size for up to four years. The roots are harvested as needed at the farmer's convenience. The plants are extremely productive.

The peeled roots of the sweet types may be prepared for eating simply by boiling or roasting. Both sweet and bitter varieties may be used to yield a coarse meal, known as *farinha de mandioca* in Brazil. The meal is often prepared by placing cut roots into a long sleevelike basket, known as *tipiti* in Brazil, which is then tied to a tree; next, pressure is exerted on the other end. The tipiti works something like the Chinese finger-locks, with the pressure extracting the juice, which is also collected and often used to prepare sauces or beers. Among some Indians in lowland, tropical South America the beer is prepared by old women sitting around a large gourd vessel who chew the roots and spit them into the gourd. The chewing initiates a breakdown of the starch into sugar, and wild yeasts then take over the production of alcohol. Visitors to a tribe are often expected to take a ritual drink of the beer, and to refuse to do so would be considered an insult.

At one time manioc was in demand in the United States in the form of tapioca. Once a popular pudding, tapioca has been largely replaced by gelatins and instant puddings. The fact that some people referred

Figure 7-5
Girl expressing hydrocyanic acid from grated manioc with a tipiti.

to tapioca pudding as "fish eyes," a fairly apt description, probably did not figure in its decline. Tapioca, as purchased for making desserts, is prepared by gentle heating, the partial cooking causing the agglutination of the manioc starch into small pellets. In addition to use as food, manioc starch is used in the manufacture of adhesives and cosmetics, for sizing textiles and in making paper. As many of its former uses have now been taken over by starch from waxy maize, little manioc now enters into international trade.

In some places, particularly in Africa, the leaves are used as a pot herb. Since the leaves may contain up to 30 percent protein, their wider use might help prevent malnutrition among manioc-root eaters. One of the aims in present manioc improvement programs, which unfortunately are being conducted only on a very small scale, has been an attempt to increase protein content. Although most varieties have

1 percent or even less protein in the roots, a few have been reported to have nearly 3 percent, although the high protein types also produce small, somewhat woody roots. Hybrids have been made in hope of transferring the character of higher protein content to more productive varieties. Attempts are also underway to produce types low in cyanogenetic glucosides, and some success has already been achieved in breeding for resistance to a virus that has caused considerable loss in Africa.

Yams

In prehistoric times the most widely distributed of the starchy crops were the yams, various species of the genus *Dioscorea*. There is no need to call upon man as an agent for their very wide dispersal, for the genus contains some 600 species, native to the tropics of both hemispheres. Man independently in many different areas discovered that the large underground stems, or tubers, were a good source of food. The tubers of some species under cultivation may reach remarkable size, 6–9 feet long and weighing more than 100 pounds.

The true yams are largely confined to the tropics and are little known in the United States. Most of the so-called yams in markets in the United States are moist-fleshed varieties of sweet potatoes. One species, *Dioscorea bulbifera*, is sometimes cultivated in greenhouses under the name aerial potato or yam potato. The small tubers or bulbils produced on the vine are sometimes used as food in parts of Asia and Africa.

Yams grow best in humid and semihumid regions. The plant is usually grown on a small scale by farmers for their own use, frequently in shifting cultivation, so that new fields are sought after a few years. Tuber cuttings, small tubers, or bulbils are used for planting. The plants are usually grown in ridges or mounds and stakes are often provided as supports for the vines. The harvest season, which is an important occasion to those people for whom this plant supplies the major food source, is celebrated with special rites. After harvest, the tubers are stored and eaten boiled, roasted, or fried as they are needed.

Today the greatest production is in West Africa, where in many places the yam is the principal food plant, as it is in parts of south-

Figure 7-6
A. Greenhouse plant of potato yam *(Dioscorea bulbifera)* showing the aerial "potatoes," or tubers. *B.* Tubers of *Dioscorea alata.* Some tubers of this yam grow to be several feet long. (Courtesy of USDA.)

eastern Asia, which is second in production. Large amounts are also still cultivated in the Caribbean. The species introduced from the Old World, which first came to America as food supplies in slave ships, are now probably more extensively cultivated in the West Indies than is the native American plant.

Yams, however, no longer are as important in the Old World as they once were, chiefly because of the introduction of other tuber crops, particularly manioc. In one sense this is unfortunate, for yams have a higher protein content than has manioc. On the other hand, the large amount of manual labor required to grow yams makes the crop a relatively inefficient one in terms of food yield for the man-hours spent on it. Little work has been done in an attempt to improve the yam, chiefly because it is a crop that is consumed locally and does not enter into trade with the developed nations. The cultivated plants are little used except as a source of food.

Some of the wild species have also been used as food in periods of famine by some people in the tropics. Many of the wild species, however, contain toxic substances and require special treatment to make them safe to eat. They have been known to cause deaths in humans. It is some of these same toxic substances, however, that have made the wild species useful to man in some other ways. Some of them have been used as fish poisons, similar to the way rotenone is used, and around the year 1940 the steroidal sapogenins in *Dioscorea* were found useful in the manufacture of cortisone and sex hormones. At one time the steroid drugs, useful in the treatment of Addison's disease, asthma, arthritis, and skin diseases, were thought to occur only in animals and their production was very expensive. As a result of the discovery of plant sources, the cost of hormones fell from $80 a gram to $2 in ten years. A still more significant use became known in 1956 when Dr. Gregory Pincus announced that a drug derived indirectly from *Dioscorea* would stop ovulation and hence prevent conception. Up to that time the steroids that prevented conception had to be taken by injection, whereas it then became possible to use oral administration. Tests with the new birth control measure in Puerto Rico and Los Angeles were successful, and the pill was on its way. Although most birth control pills are wholly synthetic today, *Dioscorea* still figures in their origin, and in this way the plant has contributed more to controlling the world hunger problem than it will ever do as a food.

Taro

Another tropical crop that feeds millions of people is taro or dasheen. Probably originally domesticated in southeastern Asia, our earliest historical record comes from China. Details about its origin have not yet been worked out but it is known to have been carried quite early from its homeland to Japan. It was also introduced into various Pacific islands apparently by Polynesians and among them is still a staple in some areas. It eventually reached Africa and was carried from there by slaves to tropical America. It was introduced into the southern United States in 1910 as a crop in soils too moist for potatoes. It never made much impact, for it couldn't be grown economically enough to compete with other root crops.

Taro, *Colocasia esculenta*, is very similar in appearance to elephant's-ear, a plant grown as an ornamental or curiosity for its extremely large heart-shaped leaves. These plants belong to the Araceae, or aroid family, which perhaps is best known to most Americans through *Philodendron*, widely grown as a house plant. Members of this family usually contain crystals of calcium oxalate in nearly all parts of the plant and these can be toxic. Anyone who has ever bitten into the tuber of Jack-in-the-pulpit, another member of this family, is familiar with the action of these crystals. The effect might be compared to biting into a pincushion with pins present. Fortunately, the calcium oxalate crystals are usually destroyed by boiling. The American counterpart of taro is yautia, that is, any of various species of *Xanthosoma*, an ancient cultivated plant, and both it and taro are cultivated in parts of lowland tropical America today.

Leaves of taro are eaten, but the part usually consumed is the underground portion, known as a corm and made up mostly of stem tissue. The corms contain about 30 percent starch, 3 percent sugar, a little more than 1 percent protein, and are fairly good sources of calcium and phosphorus. Reportedly, thousands of varieties are known, with the flesh color of the corms ranging from white to yellow and pink. A pink-fleshed variety, which is one of the favorites today, reputedly was reserved for royalty in early times in Hawaii. One of the favorite methods of use, then and now, is to make poi. Steamed corms are crushed, made into a dough, and allowed to ferment for a few days. The dough is then eaten by dipping into it with the fingers or rolling it into small balls. In Hawaii people used to eat 10–20 pounds of it a day, to which some have attributed the obesity of the Hawaiian people, a trait greatly admired among themselves.

A

B

Figure 7-7
A. Taro plants. (Courtesy of Hawaii Visitors Bureau.)
B. Taro corms. (Courtesy of USDA.)

Commercial preparation is now carried out and has largely replaced the making of it at home in Hawaii. Not all visitors to the Islands find poi an acceptable food, some comparing the taste to library paste.

The Hawaiian banquet called *luau* gets its name from the leaves of taro, which are used as part of the meal. The leaves are a good source of vitamins A and C and undoubtedly contain considerably more protein than the corms. The leaf stalks are a favorite food in much of Polynesia.

A flour, taro chips, and breakfast foods have been made from the corms in Hawaii. Since taro is easily digested, it has been recommended for use in baby foods. Today, however, outside of Hawaii there is little use of processed taro.

Taro is one of the few important cultivated plants that thrive in wet soil, although certain varieties can grow in relatively dry areas. The ancient Hawaiians accomplished some rather remarkable engineering feats to provide suitable areas for its cultivation. The propagation is vegetative, from corm tops or axillary corms, since the plant rarely flowers and seldom sets seed. The statement may sometimes be seen in older books that the plant has been in cultivation so long that it no longer flowers. Although the exact cause of the failure to flower is not known, age almost certainly has nothing to do with it. It more likely reflects a hybrid origin, mutations, or the fact that the plants are cultivated in areas where the day during the growing season is of the improper length to induce flowering.

Breadfruit

Another plant that has served as a staple, although not as important as any of the ones already discussed, is the breadfruit, *Artocarpus altilis*, a member of the Moraceae or mulberry family. The breadfruit is a handsome tree, 40–60 feet tall, with shiny deeply lobed leaves. The large fruits, which in reality are multiple fruits since they develop from the ovaries of a tight cluster of flowers rather than from a single flower, sometimes reaching a foot in diameter and ten pounds in weight, are a rich source of carbohydrates. The fruit has been used as a food in Polynesia since prehistoric times.

Captain James Cook in his voyages in the Pacific had seen the tree and from his descriptions some Englishmen thought that it would make a wonderful food for slaves in the West Indies. Captain Wil-

Figure 7-8
Breadfruit tree. (Courtesy of Hawaii Visitors Bureau.)

liam Bligh, who had sailed with Cook, was commissioned to bring trees from Tahiti to the West Indies. Thus began the now famous voyage of the H.M.S. *Bounty* in 1789. After taking on board more than 1000 young trees in Tahiti, the ship sailed, and the mutiny led by Fletcher Christian took place, with the result, of course, that the trees never reached their destination. The exact cause of the mutiny is not completely clear to this day—some have held that Captain Bligh's behavior was responsible, others, including Bligh, have thought the attractive native women may have played a role. Some of the mutineers did remain in Polynesia and married local women. Captain Bligh

Figure 7-9
Jackfruit. (Courtesy of USDA.)

and 18 faithful sailors were put on board a small boat and a month and a half later arrived safely at Timor. In 1792 Bligh remade the journey and this time did manage to transport trees to the West Indies. Such a story obviously should end with the breadfruit becoming an important food plant in the West Indies, but it never did. The West Indian blacks did not eagerly adopt it, much preferring bananas and plantains and other foods already familiar to them. But as a testimony to Captain Bligh's persistence the breadfruit is now well established in tropical America, the trees being appreciated for their ornamental value and occasionally used for food. Captain Bligh has been honored by having another tree named after him, *Blighia sapida*, the akee, whose fruit is edible, but if eaten when unripe or overly ripe can cause death.

Both seedless and seeded forms of breadfruit are known. Seedless breadfruit is generally prepared by boiling or baking. The seeded type is grown primarily for its seeds, called breadnuts, which are cooked and eaten. In parts of the Pacific area a cloth is made from the fibrous inner bark of the breadfruit tree. The details of the domestication of the breadfruit are unknown, but it has been suggested that the cultivated plant is of hybrid origin. The seedless forms obviously represent variants selected by man since they can be propagated only by him.

The genus *Artocarpus* also contains another species grown for its edible fruits and seeds, the jackfruit, *Artocarpus heterophyllus*, native to the Malay region and widely distributed in the tropics today, although of less importance than the breadfruit. It, too, is an attractive tree, differing from the breadfruit in having entire rather than lobed leaves and much larger, sweeter fruit. The fruit, which is reported at times to reach lengths of nearly three feet and weights of more than 75 pounds has been stated to be the largest fruit of any cultivated plant. It may well be the largest fruit of any cultivated tree but there are pumpkins and squashes on record that far exceed it in size and weight. Moreover, these latter fruits each develop from the ovary of a single flower whereas the fruit of the jackfruit, like that of breadfruit, is made up of the ovaries from many individual flowers.

Bananas

Many people who think of the banana only as a dessert fruit may be surprised to find it included with the staples. In many parts of the tropics, particularly in East Africa, it is the principal food of various peoples. Of the 20 million tons of bananas produced annually, only about 15 percent enters the world trade, the remainder being consumed locally. About one-half of the bananas are eaten raw in the way familiar to us, and about one-half are eaten cooked as a vegetable. The dessert, or sweet, banana is sometimes cooked but most cooking bananas are starchy rather than sweet and are referred to as plantains or cooking bananas.

The bananas had their origin in southeastern Asia, many in the Malay region. Our earliest record is an account from India in 500 BC, but it is generally assumed that the banana is a much more ancient crop, the exact age of which is unknown. Wild bananas have relatively small fruits with many hard seeds and probably were not a particularly attractive food to man. Other parts of the wild plants may have been eaten more frequently—the large underground part or corm, the shoots, and the large male bud—as they still are in some places in the tropics. The leaf stalk also may have been used by primitive man for its fibers. Thus other parts of the plant were probably much more important to man until there was a genetic change that led to seedless fruits. Man was fortunate to discover such a plant, and the plant, of course, now had to depend upon man for its spread, and there was selection for increased size and improved flavor as time went along. Seedlessness in bananas derives from both parthenocarpy, or the development of fruit without pollination, and sterility. The species that is ancestral to our domesticated bananas is *Musa acuminata*, which still exists in the form of numerous races in southeastern Asia. At some time this species hybridized with another, *Musa balbisiana;* today some of our cultivated bananas are "pure" *Musa acuminata* and others contain one or two chromosome sets from *Musa balbisiana*. An important event in the development of the edible bananas was the addition of a chromosome set to the normal diploid set. Such triploids, derived from crosses of diploids and tetraploids, are more productive and vigorous than diploid bananas and are also rather highly sterile and quite variable, giving man some superior plants to choose from. Today most of our bananas are triploids, although some diploids and a few tetraploids are cultivated.

From southeastern Asia the banana reached Africa at about the beginning of the Christian era, along with several other food plants. Some have thought that the introduction of these plants led to a population explosion in Africa at this time. The plant was first heard of in Europe from a report of Alexander the Great, and Pliny wrote that it was the plant of wise men—hence one of Linnaeus' name for the banana, *Musa sapientum,* "of the wise men." Another Linnaean name formerly used was *Musa paradisiaca,* for the plant was thought to have been regarded as the Tree of Paradise or the Tree of Knowledge among some people. From Africa the banana was carried to the

Figure 7-10
Bananas. The pointed structure at the tip of the bunch (bottom) is the male bud.
(Courtesy of USDA.)

Americas in 1516 and became so well established in a short space of time that some of the early travelers thought that it was an indigenous American plant. From Africa, too, came the name banana.

The modern history of bananas began in the last half of the past century when schooners began carrying bananas from Central America. One of the most significant events occurred in 1871 when a railroad was built in Costa Rica. Looking for something for his railroad to carry, the builder, the American magnate Minor Cooper Keith began banana plantings in that country three years later. In

1899 the Keith interests and the Boston Fruit Company merged to form the United Fruit Company, which was to become dominant in the banana industry. In the first years of this century refrigerated ships began to operate and bananas started to come to the United States with some regularity. The United States now imports more than any other country, half of the total of the world's exports, most of them coming from Central America. In the early years Honduras and Costa Rica were among the world's leading exporters, but when a fungus disease destroyed many of the commercial plantings in Central America, the small South American republic of Ecuador became the world's chief exporter, a position that it has held until recently. The Central American countries are now growing varieties resistant to disease, and although many people do not find their flavor as pleasing as that of the older variety, they are now supplying most of the export trade, and Ecuador has lost its former position. Since bananas had become the principal export crop of Ecuador, its whole economy was threatened.*

The banana not only comes in one of the neatest and most convenient packages of all of our food plants, but it is also one of our best energy sources. Its nutritive value is very similar to that of the white potato, although in the dessert varieties more of the carbohydrate is in the form of sugar. The average person would need to eat about 24 bananas a day if they were his sole source of calories. Bananas contain considerable amounts of a substance called serotonin that may be slightly poisonous, and Pirie has written that "dependence on them as the main source of energy would be inadvisable." But in parts of the world man does use them as his principal food source.

The banana plant, contrary to popular notion, is not a tree. The trunk of the banana is not woody and is not even a stem but consists of the leaf stalks. It also may surprise some people to learn that the fruit is classed as a berry. The plant is a perennial herb, the aerial portions arising from a corm. The pseudostem produces a bunch of bananas and then is removed or eventually dies naturally, but side-shoots or suckers from the same corm continue to grow and stems from these will produce bunches of bananas in the following seasons.

---

*Since some have pointed out that in the United States, cereals were stored in granaries while much of the world went hungry, it might also be mentioned that bananas have been know to rot in coastal Ecuador while some people in the highlands of the country went hungry. Similar examples could be drawn from many countries.

Suckers or corms are used for propagation. In a sense the plant is immortal, although in practice most banana plantations are started anew after 5–20 years. Some, however, have been known to remain in production up to a hundred years.

The banana is for the most part a tropical crop. It needs considerable warmth and water with adequate drainage. Some bananas, however, are produced in relatively dry, subtropical areas usually under irrigation. The plant needs little attention other than pruning to remove unwanted suckers. The plant is fairly taxing of the soil, and clean cultivation, or removal of all weeds in the plantation, which tends to promote erosion, is no longer recommended.

Only a very few of the some 300 varieties of bananas ever reach the United States and by far the most common in the past was the variety Gros Michel. The varieties known as Valery and Cavendish, more resistant to some diseases, have now largely replaced it. Other varieties are known that are better tasting but unfortunately they don't ship as well as those previously named. Bananas are cut while quite green, even for local use in the tropics where the sweet types are allowed to ripen in a shady place near the house to be used as needed. Fruit intended for export is ripened under carefully controlled conditions and ethylene is sometimes used to hasten the process.

Although more than 90 percent of the bananas grown are used directly for food, several products are made from the remainder. A banana flour or powder may be produced. In the primitive method of drying the fruits for flour, the bananas were placed in heaps on mats over a mixture of cow dung and water and covered with leaves. Modern methods involve slicing the bananas and drying them artificially. Candies or various confections are made by splitting and drying the bananas. Sliced dried bananas are used as banana chips. A beer for local consumption is made from bananas in parts of Africa. Other parts of the plant are sometimes used for food. The leaves are often employed for wrapping—the thick waxy covering of the leaf makes it an ideal "wax-paper"—or for "plates," and at times for emergency umbrellas.

The leaf fiber of the banana is of no commercial importance but other species of the genus have valuable fibers. *Musa textilis*, commonly known as abaca, or Manila hemp, which looks very much like a banana but whose fruit is inedible, produces a very strong fiber in the leaf stalk and is used to make high grade cordage. It is one of the

Figure 7-11
Harvesting abaca. The trunk, which is made up of the leaf stalks, is used for its fiber. (Courtesy USDA.)

principal exports of the Philippines, and became well established in tropical America during World War II and was grown widely in Central America on banana plantations devastated by disease.

Scientific breeding work with the banana has been carried out for only half a century, nearly all of it in Trinidad. Much of the early work was necessarily concerned with learning as much as possible about the plant. Seeds are required for breeding work, and fortunately most edible bananas will produce seeds, although in very small numbers, if pollinated. Only a start has been made in an attempt to breed new improved varieties.

8

Man's most useful tree: the coconut

No part of the coconut tree is wasted.
Malayalam proverb.

Although most people in the temperate zone are aware that palms are graceful, attractive trees of the tropics, probably few realize their great significance to man. In fact, many botanists consider the palm family (Palmae) second only to the grass family in its importance to mankind, and in many parts of the tropics palms are far more important than are the grasses. There are more than 2000 species of palms, and a long list of useful ones could be given that would certainly include the coconut, the date palm, the African oil palm, the rattan palm, the wax palm—carnauba, the world's preferred wax, comes from a palm—and the sago palm,* whose stem yields a starchy food used in Malaya and Indonesia.

The date palm *(Phoenix dactylifera)* has been considered the "tree-of-life" in the subtropical deserts of the Old World and was a symbol of fecundity and fertility. Its fruit has long served nomadic Arabs as a staple food. In addition to its high sugar content, (about

---

*A number of plants share the common name sago palm, including several different species of the palm family; some plants, however, known by this name are not members of the family.

Figure 8-1
Date palm in fruit, California.

70 percent) the date contains about 2 percent protein, 2 percent fats, and is a fair source of some vitamins and minerals; it thus is a considerably better food than are some of the starch crops. Like most palms, the date palm has many other uses—360 of them according to an ancient Persian source. Although the date palm is widely used, the coconut, *Cocos nucifera*, is rated as the world's most important palm, belonging on the list of man's dozen most important food plants.

In addition to having been called "man's most useful tree," the coconut has also been referred to as "one of Nature's greatest gifts to man" and "mankind's greatest provider in the tropics." Man uses practically all parts of the plant in one way or another, but it is the

Figure 8-2
Coconut. Left to right: Fruit as it comes from tree. With outer rind removed to show coir. Coir removed, exposing shell; this is the way coconuts usually appear in markets in the United States. Shell broken to expose the meat. Coconut water.

fruit, botanically classed as a drupe and not a nut, that gives the plant its great economic importance. The fruit is made up of a smooth outer layer, a fibrous middle layer, and a strong inner portion or shell that encloses the single seed. Frequently coconuts appearing in markets in the United States have had the outer layers of the fruit removed so that the shell is exposed. The seed is made up of an outer layer, the "meat" or kernel, and the water, which together form the reserve food for the embryo. The coconut seed is one of the largest known.*

Coconuts, like bananas, need plenty of warmth and moisture and good drainage. They are mostly found near the coast but do grow considerably inland in some regions. Florida is the farthest away from the equator where they are known to grow. Although now grown on plantations, they are still a crop on small holdings in many places. Ninety percent of coconuts are grown in southeastern Asia. The Philippines are now the largest producer and coconuts rank as the number one cash crop in that country. Indonesia, which until World

---

*The distinction of having the largest seed apparently belongs to another palm, *Lodoicea maldivica*, known as the double coconut, Seychelles nut, or *coco de mer*, which has a fruit two or three times the size of the coconut and weighing up to forty pounds. Marvelous tales were once told of the fruits that washed up on the shores of India from their homeland in the Seychelles Islands.

War II held the first position, is now second in production, followed by Ceylon, Malaya, and Mexico.

Coconut trees have been reported to bear 500 nuts a year, but 50–100 seems to be a more normal production. Harvesting is usually done by climbing the trees or by cutting the nuts with knives attached to long bamboo poles, or more rarely the nuts are allowed to fall to the ground and are then collected. Monkeys reportedly have been trained to harvest coconuts in Sarawak, Indonesia, and Thailand. The name *Cocos* itself relates to monkeys but has nothing to do with their harvesting the fruit. The word, which comes from the Portuguese, means monkey's face, in reference to the three eyes in the shell that make it resemble the face of a monkey.

After the nuts are harvested, they are cut in half and the meat gouged out immediately or after partial drying in the sun. The meat is then cured by sun drying where weather permits or in kilns to produce copra. The moisture content must be drastically reduced or the copra deteriorates rapidly. After drying, the oil, which forms 60–70 percent of the copra, is extracted. Primitive methods using stone or wooden mortars and pestles, powered by humans or bullocks, are still employed in parts of India but have been replaced by hydraulic presses in most other coconut-producing areas. The oil is the most important commercial product derived from the coconut. Its greatest use in the last century was in the manufacture of soaps. All floating soaps were made from coconut oil until fairly recently, when it was found that soaps made from other oils would float if air was pumped through them during manufacture. Coconut oil is still regarded as one of the best for making soaps and considerable amounts are still used for that purpose. In this century, however, its chief use has been for margarine, and it was the main oil used for that purpose until recently when soybean and cottonseed oils have come largely to replace it. Since coconut oil is composed of 90 percent saturated fatty acids its use is not as highly recommended for human food as the other vegetable oils, which are largely unsaturated.

The residue, or coconut cake, left after oil extraction, is used chiefly to feed livestock. The coconut cake is a rich source of both protein, one of the most nearly complete proteins of all vegetable sources, and carbohydrates. Although it is more fibrous than most other oil-seed residues and hence difficult for man to digest, it is nevertheless unfortunate that more of it is not processed for human

Figure 8-3
A. Coconut tree in fruit. (Courtesy of FAO.)
B. Monkey harvesting coconut. (Courtesy of Sa-korn Trinandwan.)
C. Coconut "monkey face."

food, for coconuts are mostly grown in areas where protein deficiency is often pronounced.

The meat, from either immature or mature coconuts, is, of course, used directly for human food in many areas where it is grown. Although it is the chief vegetable protein source for some people, it is usually served mixed with other food as a vegetable and nowhere does it appear to be the basic food staple as do most of the plants previously discussed. The per capita consumption is estimated to be 140 nuts a year in Ceylon, and is probably considerably greater than this among Polynesians in some of the Pacific Islands. In many countries of the temperate zones, dried coconut is used mostly for candies and cakes. Desiccated coconut meat was first made in England and the United States in the last part of the past century and is a fairly important use of coconut meat today.

The fibrous part of the fruit or husk, known as coir, also has a number of uses. It makes a fine rope, resistant to sea water, and its use for this purpose is quite ancient. Coir is also used to make mats, rugs, filters, and for stuffing furniture. India is the world's greatest producer. To prepare the fibers, the husks are immersed in saline backwaters for several months for retting and the fibers are then separated by beating the husks with wooden mallets or clubs.

The young inflorescence, or flower cluster, of the coconut, like that of several other palms, yields a sweet juice, or toddy, when tapped. The toddy, which is mostly sucrose, is drunk fresh or, more frequently, is used to prepare alcoholic beverages, such as arrack, or vinegar. In Ceylon more than eight million gallons of arrack are produced annually, very little of it being exported. Small amounts of toddy are also used for making sugar.

The shells have a number of uses, the most ancient of which is for eating or drinking utensils and for fuel. They are still used for these purposes as well as for bowls for hookah pipes, for the manufacture of novelties or "artistic objects," and, when ground, are used as a filler in plastics. The coconut water makes a refreshing drink, and in recent years has been used by plant physiologists as a growth-promoting substance. The large leaves, which reach lengths of twelve feet, are used for thatching, and for making baskets and hats.* The wood is used to some extent in construction and furniture making

---

*They are apparently second only to the leaves of the Panama hat palm for this purpose. The Panama hat palm isn't a true palm but rather belongs to the family Cyclanthaceae.

Figure 8-4
Extracting coconut meat for copra, Caroline Islands. (Courtesy of S. F. Glassman.)

and forms some of the "porcupine wood" of commerce. The palm heart, or cabbage, the tender bud at the apex of the stem, is sometimes eaten, but not as frequently as that of some other palms, for once the bud is removed the tree dies. The coconut is also still used as a religious offering in some parts of southeastern Asia, probably stemming from the ancient belief that the coconut is the "Tree of Heaven," *Kalpa Vrikska* in India, or the "Tree of Life." There is also a belief among certain natives of New Guinea that the first coconut tree sprang from the head of the first man to die.

Two main groups of coconut palms are recognized—dwarf forms and tall forms. Numerous varieties exist within each group, differing primarily in the shape, size, coloring, and yield of the fruit. Although seed selection for high yield probably is fairly ancient, there has been very little modern scientific breeding work on the coconut. There are several reasons for this. First of all, it is a tropical crop and, as we have already seen for other plants, these have received much less attention than the crops of the temperate zone, where more plant breeders and money for research are available. Moreover, the coconut is often a small landholder's crop rather than a plantation crop. As a rule governments seldom do as much for the small farmers as they do for the big ones. Nor can small landholders employ breeders as do some of the companies that have extensive holdings. Another factor responsible for the limited improvement of the coconut is that breeding work with a tree always takes longer than that with annual or herbaceous perennial plants. It takes three years for a dwarf coconut palm to bear fruit and five to seven years for the tall varieties. From time of flowering to fruit maturity is nearly a year. Germination of the seed requires about four months, time enough for a full growth cycle in some annual crops. Thus the breeding of superior varieties of the coconut through hybridization requires many years of work.

The place of origin of the coconut has been the subject of some controversy in the past. There were some who held that the coconut was a native of the New World, primarily because all of its close relatives are American. Others maintained that the coconut was native to the Indo-Pacific region, pointing out that it was a much more extensively used plant in this area than in the Americas. While it is true that the coconut does have most of its close relatives among the American palms, there is now general agreement among botanists that it originated in the Indo-Pacific region. Among some of the most convincing evidence that has been brought forward since the original controversy is fossil *Cocos* of late Tertiary age in New Zealand and India, proving that a species of coconut did inhabit the Pacific region before modern man appeared on the scene. The coconut was present in both Asia and the Americas previous to 1492. At that time, however, it was known only from western Panama in the Americas and its wide distribution in the New World came in historical times. Its presence in Panama previous to the arrival of the Spanish can probably be explained without invoking man's aid, for coconuts are known to be able to float in sea water for more than 100 days, which would allow ample time for a fruit to float across the Pacific. Probably the establishment of coconuts in new areas through the agency of ocean currents is a rare event, but it would need to have happened only once to explain the plant's presence in the New World. Man, as well as ocean currents, is probably responsible for its wide distribution on the Pacific Islands. How early the coconut first became a domesticated plant is not known. It was in India by 1000 BC but it may not have been first domesticated there.

Two tales concerning coconuts, widely circulated and even found in some recent textbooks, were not put to rest until recently. One is the story of the coconut crab or robber crab. According to the account given by Darwin, this crab tears the husk from the coconut "fibre by fibre" exposing the three eyes on the shell and then hammers on one of the eyes with its heavy claw until a hole is made, after which it extracts the coconut meat with its pincers. Apparently Darwin's leg was being pulled by a Mr. Liesh, whom he credits as his source, for recent studies have indicated that no one has ever seen a crab in nature perform this remarkable feat and in feeding experiments, crabs have died when coconuts were the only food offered them. The second story, concerning the finding of pearls inside of coconuts, also has never been substantiated. The so-called coconut pearls in museums have been shown to come from molluscs.

9

Other plants for food, beverage, and spice

Cauliflower is nothing but cabbage with a college education.
MARK TWAIN, *Pudd'nhead Wilson's Calendar.*

Although man can, and sometimes does, live solely on the basic foods, he usually uses other plants as food or beverage with every meal. Some of these other foods are eaten primarily to add variety to the diet and to increase the enjoyment of eating, but at the same time many of them are excellent sources of vitamins and minerals as well as supplying small amounts of carbohydrates, proteins, and fats. Primitive man, as pointed out in the first chapter, almost certainly exploited a great array of wild food plants and he continued to use some wild plants as fruits and vegetables after the domestication of the cereals. Eventually some of these fruits and vegetables were cultivated and in time became completely domesticated species. Man today still utilizes wild food sources in many parts of the world, often out of necessity but sometimes by preference. The collecting of wild foods has become an interesting hobby for many people in the United States.*

Vegetables

Vegetable has no precise botanical meaning in reference to food plants, and we find that almost all parts of plants have been employed as vegetables—roots (carrot and beet), stems (Irish potato and asparagus), leaves (spinach and lettuce), leaf stalk (celery and Swiss chard), bracts (globe artichoke), flower stalks and buds (broccoli and cauliflower), fruits (tomato and squash), seeds (beans), and even the petals (*Yucca* and pumpkin). A great many different plant families have provided our vegetables, but just as certain families have been particularly important in giving us our staples, so too have certain others in giving us vegetables.

The mustard family, or Cruciferae, has been particularly significant for its vegetables, and a single species, *Brassica oleracea*, has provided us with the cole crops, which include cabbage, kale, Brussels sprouts, cauliflower, broccoli, and kohlrabi. The kohlrabi, although often cultivated in the United States, is seldom seen in markets. In this variety the stem enlarges above ground level to produce an edible tuber. The ancestor of all the cole crops was native to the Mediterranean region, and some think that it was first used for its oily seeds. Selection with emphasis on different parts of the plant eventually produced the diversity of cultivated varieties now enjoyed by man.

Other species of *Brassica*, originally native to either Europe or Asia, account for a number of other vegetables. These include plants grown for their roots, the turnip and rutabaga, and a great number whose leaves are eaten and collectively are designated as the mustards. One of these, black mustard, has seeds that are used in the preparation of the mustard of our tables and widely used as a condiment. Mustard is also used in medicine as a rubefacient, or counterirritant.

---

*Many of the "wild" foods being gathered in the United States are introduced species that have escaped from cultivation rather than native wild plants. Wild asparagus, wild carrots (also known as Queen Anne's lace), prickly lettuce, chicory, and burdock, for example, are plants that were introduced, either intentionally or accidentally, from the Old World and have become widespread weeds in North America. The Jerusalem artichoke and cattail, also used as wild food sources, however, are native North American species.

Wild type

Kale

Cabbage

Cauliflower

Brussels sprouts

Broccoli

Kohlrabi

Figure 9-1
Variation in *Brassica oleracea*. By selection man has produced varieties for their edible leaves (kale and cabbage), specialized buds (Brussels sprouts), flowering shoots (cauliflower and broccoli), and enlarged stems (kohlrabi).

The radish *(Raphanus sativus)* is another Old World member of the mustard family employed for food. Although radishes are ordinarily used only as a salad ingredient or relish in many places, they have an important role as a food plant in the Orient. Japanese varieties may reach weights of 65 pounds. They are used as a cooked vegetable, often stored for use in the winter, and are also fed to livestock. In parts of Asia one variety of radish is especially grown for its seed pods, which may reach lengths of two feet and are used as a vegetable.

Of equal or greater importance for its contributions to our vegetables, is the cucurbit family (Cucurbitaceae). Five different species of squash or pumpkin, belonging to the genus *Cucurbita*, were domesticated in the Americas. Some of these rank among the oldest known foods of the Americas, being recorded in archaeological deposits dating back to 7000 BC in Mexico. Since the wild cucurbits have little or no flesh in the fruit, it has been postulated that they may have been domesticated for their edible seeds. ("Pepitas" or pumpkin seeds are, of course, still eaten.) Mutant types with fleshy fruits then appeared, according to the theory, and selection by man has produced the thick-fleshed varieties now widely cultivated. The squashes and pumpkins, along with maize and beans, were carried north from Mexico and became the principal food plants of the North American agricultural Indians. Following the discovery of America, pumpkins and squashes were soon introduced to Europe and Asia, and today are important in many parts of the world, not only for human food but for livestock as well.

The Old World has also furnished food plants from the Cucurbitaceae, including the cucumber, melons, such as cantaloupe, cassaba, and watermelon. The cucumber and the melons come from different species of the genus *Cucumis;* the watermelon belongs to the genus *Citrullus.*

The bottle gourd *(Lagenaria siceraria)*, another member of the Cucurbitaceae, although never more than a minor food plant, was particularly valuable to man for its hard-shelled fruits which were used as containers, for musical instruments, for floats, as well as in other ways. This species is thought to be native to Africa, and archaeological remains of the fruit have been found in Peru dated at before 10,000 BC, and in both Mexico and Thailand dated at about 7000 BC. Some people have thought that man may have been responsible for its wide dispersal in early times, but since it has been shown that the gourds can remain in sea water for long periods without damage to the seeds, it is perhaps more likely that its wide distribution is to be explained as the result of oceanic drifting of the fruits. It was probably man's most widely distributed domesticated species in prehistoric times and continues to be fairly widely used throughout much of the tropics.

The nightshade family, or Solanaceae, in addition to providing the Irish potato, has supplied us with several other food plants, the most important of which is the tomato *(Lycopersicon esculentum)*. The tomato was already a well established cultivated plant in Mexico when the Spanish arrived. It reached Europe in the first half of the sixteenth century, and somehow acquired the reputation of being harmful to eat, the reasons for which are not entirely clear. It seems likely that the tomato was recognized as a member of the nightshade family, known to Europeans of the time as comprising only poisonous plants, such as deadly nightshade, henbane, and mandrake, and hence people were reluctant to eat it. Tomatoes are known with yellow, orange, pink and green fruits in addition to the familiar red types. One of the first tomatoes to reach Italy was a yellow-fruited variety called *pomi d'oro* (apple of gold), which somehow became transformed to *poma amoris* (apple of love). The name love apple soon became attached to it, not because of any real or supposed aphrodisiac property, but simply through the translation of the transformed Italian name. Only in this century did the tomato finally become widely appreciated for the fine food that it is. Remarkable achievements have been accomplished by the plant breeder in its improvement in recent years, one of which is the development of forms with special characteristics that allow it to be mechanically harvested.

Among other food plants in the Solanaceae are the eggplant and the sweet and hot, or red, peppers. The eggplant, a species of *Solanum*, apparently had its origin in India, and, like the tomato, it was regarded with suspicion when it first reached Europe. One name for it at that time was mad apple, for it was thought that the eating of it would produce insanity. Several different species of *Capsicum* were domesticated in tropical America for their pungent fruits and these peppers became almost indispensable in the diet of many Indians. In post-Columbian times they became widely dispersed by man and have become as important in parts of southeastern Asia and Africa as they are in their homeland. Sweet peppers, a variety of *Capsicum annuum*, which also includes cayenne and chili peppers, have become more important in the temperate zones than have the pungent forms.

Figure 9-2
Tomatoes that are tough
skinned, even ripening,
readily detachable, and of
an uniform size have been
developed for machine
harvesting. Mechanical
harvester seen in the
background.
(Courtesy of USDA.)

Figure 9-3A
Sugar beet.
(Courtesy of USDA.)

Spinach *(Spinacia oleracea)*, beets *(Beta vulgaris)*, and the pseudo-cereal quinoa *(Chenopodium quinoa)*, are all members of the Chenopodiaceae, or goosefoot family. Leaves of several wild members of this family have long been used as potherbs by man. Spinach of southwestern Asia became a domesticate and one of the world's important leafy vegetables. Man also used the leaves of a wild beet of coastal Europe for food and it, too, became domesticated. Some varieties of beet, such as Swiss chard, are still cultivated as leafy vegetables. The enlarged root of the beet is apparently a result of selection by man following its cultivation as a leafy vegetable; ultimately the root gave the plant its greatest significance—not as a vegetable but as a source of sugar. The recognition in the latter part of the eighteenth century that the beet contained sugar was followed by selection and breeding work that has increased the sugar content from about 2 percent to 20 percent. Napoleon, in the early part of the nineteenth century, in spite of ridicule, encouraged the new sugar-beet industry, realizing that it could give France a domestic source of sugar and free

Figure 9-3B
Quinoa in the highlands of Ecuador. This plant is an important source of food in much of the high Andes.

it from dependence on England for sugar, which had a monopoly on sugar cane. The sugar beet has since become a major crop in the north temperate zone. Sugar-beet growing became established in the United States in the latter part of the last century, and has since become important in several of the western states. Mechanization of nearly all of the operations connected with sugar-beet growing and harvesting has allowed sugar from beets to compete favorably with that from sugar cane, a tropical crop whose production still employs much cheap hand labor. Quinoa, whose leaves are sometimes used as a vegetable, is cultivated for its seeds in the very high Andes where it is an important food source. The seeds are used in thickening soups, making a porridge, and sometimes to make a chicha. Since quinoa is a good protein source, it is very important in an area where the Irish potato and other starchy crops predominate among the food plants.

Fruits

Fruits, wild or cultivated, must have always been a source of pleasure for man because of their sweetness. Like vegetables, they come from many different families of plants, but in the north temperate zone, one family—the rose family, or Rosaceae—stands out. Among its more important contributions are the pome fruits, the apple and the pear, species native to West Europe and Asia; the stone fruits, various species of the genus *Prunus*, which include the peach, cherry, plum, and apricot, most of which also come from the Old World, although species of cherry and plum were also domesticated in the Americas; and the "berry"* fruits, blackberry, raspberry, and strawberry, with species from both the New and Old Worlds entering into domestication. Although these are rightly called temperate-zone fruits, some of them are cultivated at high altitudes near the equator.

Tropical and subtropical fruits abound, but only a small number such as bananas, pineapple, figs, and the citrus fruits reach the temperate-zone markets with any regularity. The citrus fruits, whose nutritional value is widely recognized, are all members of the family Rutaceae, and a single genus, *Citrus*, supplies us with the most important species—sweet, bitter or sour, and mandarin orange (one form of which is known as the tangerine in the United States), lemon, lime, grapefruit, citron, and the shaddock, or pumelo. All of these originated in southeastern Asia with the exception of the grapefruit, which is thought to have been derived from the shaddock after it was introduced in the West Indies. The citrus plant is a small tree, often somewhat spiny, and having attractive fragrant flowers. The fruit is a special type of berry known as a hesperidium. Its thick leathery rind bears numerous oil glands that yield an essential oil widely used in flavoring.

Scientific proof of the importance of the citrus fruits came in 1756

---

*Most of the so-called berry fruits do not have fruits that are classified as berries according to the botanical definition. A true berry is defined as a fleshy, many-seeded, indehiscent fruit developing from the ovary of a single flower. Examples are grape, tomato, pumpkin, and orange. The raspberry and blackberry are in reality aggregate fruits in that they develop from many ovaries of a single flower. The individual fruitlets or seed-bearing structures of the blackberry, for example, each are the product of a single ovary and are the equivalent of a plum in that there is a single seed enclosed in a fleshy covering. The strawberry is defined as an accessory fruit, for the fleshy part develops from a structure other than the ovary. The true fruits of a strawberry are the small, hard, straw colored "seeds" on the surface.

when John Lind, a surgeon in the English navy, found that scurvy, particularly common among seamen at the time, could be prevented by eating oranges and lemons. Later in the century the Royal Navy began to provide rations of lime or lemon juice to its men, and the name "limey" came into use for British sailors as a result. Not until 1933 was vitamin C (ascorbic acid) identified as the factor responsible for the prevention of scurvy.

Nuts

Nuts of various kinds, in addition to the peanut and coconut, which were previously discussed, are a highly concentrated source of food and have long served man as an important energy source. From the archaeological record we know that nuts of various wild plants were a frequent source of food for prehistoric man. The word nut, as popularly used, is applied to the fruit or seed of a great number of plants, mostly trees. Botanically a nut is defined as a hard and indehiscent one-seeded fruit. Thus, of the nuts utilized by man, only a few, such as the acorn, the chestnut, and the hazelnut, meet the botanical definition. Acorns from various species of oaks, in both the Old and New Worlds, were at one time an important source of food. Many of the Indian tribes of the west coast of North America relied on acorns as their principal food source, and devised various ways of leaching the tannins and bitter principles from them in order to make them palatable. Acorns are still sometimes used as a food source by the poorer people in some of the Mediterranean countries of Europe. The Eurasian chestnut continues to be a food plant in South Europe, but the native American chestnut, whose nuts were once widely sought, has been practically eliminated by a blight disease that swept through the eastern United States at the beginning of this century. Acorns and chestnuts are valued as foods because of their high carbohydrate content.

Among the nuts with particularly high protein content are the almond and pistachio, both old cultivated plants of the Mediterranean region. The almond belongs to the same genus as the stone fruits, but the fleshy covering is poorly developed and the seed is, of course, the only part eaten. The almonds produced in the United States are grown in California.

Nuts with high oil content include the Brazil nut and the cashew, both native to Brazil, the pecan of the central and southern United States, walnuts, and hazelnuts. The walnut foremost in use for food comes from the English or Persian walnut, which originally came from Iran. California is today one of the leading areas for walnut production. The native American black walnut is more valued for its wood than for its nuts. The hazelnut, or filbert, of Europe is also an important yielder of nuts. There are also native American species of hazelnuts.

Beverages

Man early discovered that parts of certain plants had a pleasant stimulating effect* on him, and today many of these plants serve as man's chief sources of nonalcoholic beverages. The four most significant ones are coffee, tea, cacao or chocolate, and maté or Paraguay tea. These plants, all members of different botanical families, share one important feature in common, the possession of caffeine or a very similar alkaloid that is responsible for their stimulating property. Except for cacao these plants offer man little or nothing in the way of nutrition, and hence are hardly essential. Even though coffee and tea are not, strictly speaking, food plants, they are extremely important export crops in many parts of the tropical world and figure prominently in the economic welfare of many countries.

From a commercial standpoint coffee is the world's foremost beverage plant, although more people probably drink tea. Native to Abyssinia, coffee was carried to Arabia over 500 years ago and for two centuries Arabia was the principal producer. Coffee was later found to be well adapted to many parts of the American tropics from elevations near sea level to 6000 feet, and today Brazil leads in the world's production. The United States is the world's chief importer of coffee, but per capita consumption is said to be greater in Sweden. The coffee plant, *Coffea arabica*, is a small tree or shrub with shiny dark green leaves and numerous clusters of white flowers. Each of the berries contains two seeds called coffee beans. Following harvest, a process of fermentation and roasting is required before the beans assume their distinctive odor and flavor.

---

*Exactly how man discovered the effect of certain of these plants is somewhat of a puzzle, for some of them have little or no stimulating action unless they are specially cured or processed.

Figure 9-4
Coffee berries. (Courtesy of FAO.)

Tea, *Camellia sinensis,* also a small tree or shrub, is indigenous to India and China where most of the world's production is concentrated today. The young leaves are carefully collected and then sorted to yield the various grades of tea. The post-harvesting processing is responsible for producing the different flavors of tea. Green teas are produced by drying and rolling the leaves, and black teas result from a fermentation during the drying process. It probably will come as no surprise that Great Britain is the world's greatest importer of tea.

Chocolate and cocoa come from the cacao plant, *Theobroma cacao*

Figure 9-5 A
Cacao tree with pods. (Courtesy of Jorge Soria.)

(theobroma, from the Greek, means "food of the Gods"), native to lowland tropical America, and apparently first domesticated in Mexico. Details of its origin, like that of many of our domesticated plants, remain to be elucidated. When the Spanish reached Mexico, they found chocolate to be a prized drink among the Aztecs. Cacao beans, in fact, were once considered so valuable that they served as currency. The cacao plant is a small tree with rather large leaves, and is rather unusual in that it bears its small flowers, and eventually pods, close to the trunk and branches. The pod, a specialized berry, yields several rather large seeds or "beans." Following fermentation (a rather odorous process), drying, and roasting, the seeds are ready to be ground. The whole bean gives us chocolate, which is a rich food as well as drink, for it contains about 30–50 percent oil, 15 percent starch, and 15 percent protein. Cocoa is produced by removing most of the fatty oils, which then are used as cocoa butter. West African countries have replaced tropical America, where diseases have al-

Figure 9-5B
Opened cacao pod, exposing
the individual seeds, or beans.
(Courtesy of USDA.)

ways plagued the trees, as the world's principal producer and now account for 75 percent of the world's production. *Cola nitida*, whose seeds are the source of cola, which is widely used in soft drinks, native to tropical West Africa, belongs to the same family as cacao. Much of the cola supply used in the United States is from trees cultivated in Jamaica. Caffeine from other sources is also used in some soft drinks.

Maté, or Paraguay tea, although less widely known than the previously discussed beverage plants, can hardly be said to be a minor beverage, since it is drunk by more than 20 million people in South America and some is exported to Europe and North America. Maté comes from the leaves of *Ilex paraguariensis*, a relative of the holly tree, and is cultivated in Brazil, Paraguay, and Argentina. The processing of the leaves and the preparation of the drink are somewhat similar to the methods used for tea. Traditionally in South America maté is drunk from a gourd cup through a metal straw.

Alcoholic beverages have already been mentioned in various places in this book, and it should be quite obvious by now that many different plants can be employed to prepare alcoholic beverages. Few, however, can rival the grape. Grapes are, of course, one of our widely used fruits, but the principal reason for their cultivation, in times past as well as today, has been for winemaking. The grape native to the Near East or surrounding area, *Vitis vinifera*, although not one of our earliest domesticates, is of considerable antiquity as a cultivated plant. Presumably, the origin of wine, which is simply fermented grape juice, was not very complicated. Someone squeezed some grape juice, let it stand, and wild yeasts converted some of the sugar in the grape to alcohol. Its preparation through the years, however, has become more elaborate. The Bible tells us that Noah planted a vineyard, and in fact, wine is mentioned no less than 165 times in the Bible. Wine had probably been around for sometime before it reached the Greeks and Romans who made considerable improvement in the "art" of wine making. The grape vine was carried to France in 600 BC, and after Christianity became established, monasteries were to play a significant role in establishing some of the great vineyards there. France was to become the world's foremost wine country—not only in consumption but usually in quality and quantity produced as well. In the latter half of the eighteenth century there were devastating losses of the vines throughout Europe from diseases, and not until it was found that stems of *Vitis vinifera* could be grafted to rootstocks of the American species was there a recovery. It was perhaps only fitting that the resistance to the diseases should come from the American rootstocks for the diseases came to Europe with the introduction of American vines in the first place. This was also the time of Louis Pasteur's great discoveries, some of which contributed directly to an improvement of the wine industry.

Attempts were made to grow *Vitis vinifera* in eastern North America soon after it was settled but both the humid climate and the cold winters were unfavorable, and it was to be some time before the people turned to the American species and achieved success. In 1852 the Concord grape (named for the town in Massachusetts) arose, either as a mutant of the native fox grape or as a hybrid with *Vitis vinifera*. It was an immediate success and its cultivation spread halfway across the continent in two years. The Concord and other native American grapes are the source of the wines of the Great Lakes region today. The Old World species, however, was very successful in the Mediterranean-like climate of California where it was introduced by the Spanish late in the eighteenth century, and today is the basis of the California wines. *Vitis vinifera* is also fairly widely cultivated in many other parts of the world for the production of wines.

A

Figure 9-6
A. California vineyard in winter. (Courtesy of Wine Institute.)
B. Harvesting the grapes. (Courtesy of Wine Institute.)

B

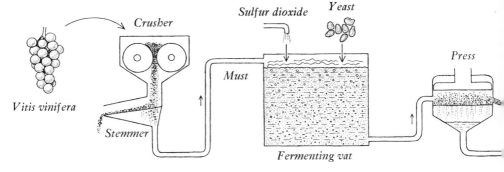

Figure 9-7
The method of producing wine in California is similar to that of Europe. Shown here are the steps in the production of red wine. The grapes are crushed to produce a "must." The must goes to a fermenting vat where yeasts transform the sugar into alcohol and then to a press for the removal of the skin and seeds. The wine then moves to settling vats where a "fining" process removes impurities. After filtering and aging in casks, it is ready for bottling. (From "Wine" by Maynard A. Amerine. Copyright © 1964 by Scientific American, Inc. All rights reserved.)

Spices

In his search for edible plants primitive man must have discovered many of those plants that now supply us with spices and condiments, and he probably learned to use these aromatic plants to make his food more flavorable or to help cover up the taste and odor of food that had passed its prime. In time some of the spice plants became intentionally cultivated, and with the Romans the spices came into their own in the art of cooking. Through Marco Polo's accounts Europeans became aware of the wealth of the spices of the Far East, and it was partly an attempt to secure these spices that led to great ocean voyages of discovery in the fifteenth and sixteenth centuries.

Though hardly essential to man's welfare, spices often made eating more enjoyable and a large number of spice plants are found among our domesticated species. Reference has already been made to red pepper and mustard. The botanical families that have made the greatest number of contributions to our spices are the mint family, or Labiatae, and the parsley family, or Umbelliferae. The former has given us basil, marjoram, oregano, rosemary, sage, savory, thyme, as

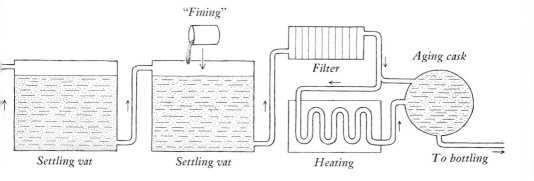

"Fining"

Aging cask

Filter

Settling vat   Settling vat   Heating   To bottling

well as spearmint and peppermint; the latter has contributed caraway, celery, chervil, coriander, cumin, dill, fennel, and, of course, parsley. Since the parsley family also provides several vegetables, such as carrot and parsnip, it must be considered one of the families of major importance to man. The orchid family, which in point of number of species is the largest in the plant kingdom, has supplied only one plant that graces our food. Extract of vanilla comes from the pod of the vanilla orchid, native to tropical America. Although the plant is still cultivated for its use as a flavoring, today most vanilla flavoring is made synthetically.

By far the world's most important spice is pepper *(Piper nigrum)*, which at present accounts for one-fourth of the world's commerce in spices. The pepper plant is a member of the family Piperaceae and is not at all related to the *Capsicum* peppers, previously mentioned. Apparently first domesticated in India, pepper became one of the first trade items between Europe and the Far East. Like many spices of the time it was widely used in medicine as well as in seasoning and preserving food. The pepper plant is a woody vine, climbing to heights of 30 feet or more. The inconspicuous clusters of flowers each produce 50 or 60 fruits known as peppercorns. After drying, the whole peppercorns are ground to produce black pepper. If the hull is first removed the ground product results in white pepper. The plant is adapted to hot, wet, tropical regions and today the bulk of the world's supply is produced in India and Indonesia. The United States is the chief importer.

Another spice, nutmeg, has received some notoriety as a hallucinogen in recent years. Its use for such purpose is actually quite ancient.

It was early recognized that overindulgence could be lethal. Nutmeg comes from a tree, *Myristica fragrans*, native to the East Indian Archipelago and now cultivated in the West Indies as well as in its homeland. The pulverized seeds produce the spice nutmeg, and a second spice, mace, is prepared from the outer growth of the seed.

There are, of course, many other plants used for food. A few of the well known and widely used ones, as well as many lesser known, have not been included in this survey. The present account should, however, give some indication of the great diversity of plants that serve man in his diet.

10

Evolution and breeding of man's domesticates

And he gave it for his opinion, that whoever could make two ears of corn, or two blades of grass, to grow upon a spot of ground where only one grew before, would deserve better of mankind, and do more essential service to his country, than the whole race of politicians put together.
JONATHAN SWIFT, *Gulliver's Travels.*

With the domestication of plants and animals came changes in these organisms under man's influence, at first largely unintentional. Today man is directing change in his domesticated plants and animals in an effort to make them serve him more effectively. Modern evolutionary theory provides an understanding of how the changes occur.

The heredity of a plant or animal is controlled by genes. Although genes are ordinarily very stable and copy themselves exactly generation after generation, spontaneous changes in them do occur. Such changes are known as gene mutations and may be passed on to the following generations. Through sexual reproduction genes may be recombined in various ways, and thus it is possible for a species to "try out" various combinations of genes. Together mutation and recombination provide for variation, the raw material for evolution.

The force that directs this variation is natural selection. Natural selection operates through differential reproduction. Thus an organism with a particular combination of genes may produce more offspring than those with other combinations. In time the more successful combinations will replace those less successful.

Once a man brings plants or animals under his care the same evolutionary factors are at work, mutations and recombinations producing variability, and selection serving to guide the changes. However, man instead of nature is the important selective agent. Thus in the process of domestication we find that artificial selection joined natural selection. The plants and animals in turn became dependent upon man for their perpetuation and frequently lost the ability to survive under conditions in nature. We have seen in previous chapters that many cereals and legumes have lost the wild-type characters for seed dispersal and germination that allow them to compete in nature and have acquired characters, such as nondehiscent fruits and large seeds, that have made them better plants for man's purposes.

Although the archaeological record has revealed a great deal about man's early foods, it hasn't as yet told how long the process of domestication took. Nor, perhaps, should it be expected to, for the process is such a gradual one that sharp demarcation between wild species and early domesticated or semidomesticated species is difficult. Domesticated plants and animals usually differ in several ways from wild ones and the differences are not acquired all at once. A species may, of course, continue to change after domestication is completed, as are many of our domesticates today. Domestication might be said to be completed when man controls the breeding of the organism, as was mentioned earlier.

The changes accompanying domestication could have occurred very rapidly in some organisms. The process probably did not require thousands of years as implied by some writers on the subject. In fact, a few hundred years or less could have been enough time for some rather profound changes in a plant or animal under man's influence. The time required would, of course, vary considerably from one organism to another and according to the type of selection that was exercised by man. Certainly annual plants, which produce a new generation every year, could have undergone a very rapid change in the hands of man.

The great Swiss botanist A. P. de Candolle pointed out a hundred years ago that no new basic food plants have been domesticated in

Figure 10-1
Domesticated and wild *Capsicum* peppers. In addition to being much larger, the domesticated peppers remain on the plant unlike the fruits of the wild type, which are deciduous at maturity.

historical times. This statement is still true and applies to animals as well as plants. How could primitive man have made such wise choices in his selection of plants and animals to domesticate? Although luck or chance may have played a role, I think that we must give the major credit to experimentation of a trial-and-error sort. Man must have had an intimate acquaintance with all the food resources in his environment when he began the process of domestication. We don't know if he started his trials with the plants and animals that were difficult to obtain or with those that were common. It would seem reasonable that he started with those that he regarded as important— whether for religious reasons or because they were his favorite foods. Some of his experiments would have failed, and some of the early domesticates may have been replaced by others that were superior for one reason or another. The archaeological record wouldn't neces- sarily be expected to give testimony of his failures. Interesting evi- dence does exist, however, that certain North American Indians had a marsh elder *(Iva annua)*, a relative of the ragweeds, that had fruits much larger than those of wild marsh elders today. It seems reasonable that these fruits, some of which are found stored in prehistoric sites, were used for food, but the plant apparently was replaced by superior foods before observations were made in historical times.

Among man's most important plants were annuals with large or numerous seeds. Many of them were probably originally weeds that grew in open or disturbed habitats. In a sense they were preadapted to cultivation since they were not only prolific but also could mature seeds in a space of a few months and were well suited to grow in man-made environments. It was, therefore, not entirely accidental that they gave rise to man's most important food plants.

In an earlier chapter it was pointed out that certain animals, cattle, sheep and goats, for example, were preadapted to man's use since they could live on a diet that did not place them in competition with him. Moreover, the fact that they were gregarious, rather than solitary, animals would mean that they could be more readily kept and managed by man. As with plants it is quite possible that attempts were made to domesticate other animals. We know that the ancient Egyptians did so.

The earliest type of selection under man's agency was probably not a conscious one. The loss of natural means of seed dispersal may serve as an example. When man brought cereal into cultivation the plants had a brittle fruiting stalk that would shatter readily, allowing some of the grains to fall to the ground before man could harvest them. A mutant type that had a nonbrittle fruiting stalk might appear occasionally and would hold all its grain until harvested. Among seeds saved for planting there would likely be those from the plants with nonbrittle stalks. After the next year's sowing and harvest more seeds from the mutant strain would be collected. This process, if repeated year after year, would lead to more and more of the plants with the nonbrittle stalks appearing in the next generations. For the nonbrittle type completely to replace the wild type considerable time would be required and would depend to a large degree upon the nature of the gene or genes controlling this particular character. If, however, man realized the advantage of the nonbrittle character and saved only seeds of this type for sowing, the character might become established fairly rapidly.

In a like manner there would have been changes in animals. For example, when he was capturing animals, man would have found it difficult to take the most vicious and cunning. Or, if such animals

A

Figure 10-2
A. Detail from Albrecht Dürer's, "The prodigal son amid the swine" (1496).
B. Present-day sows with young. Breeders have considerably changed the appearance and productivity of the pig. (Courtesy of USDA.)

B

were captured, they might have escaped or quickly ended up in the pot as man realized they couldn't be managed. Thus there would have been selection for the more docile animals to furnish stock for the next generation.

How early intentional selection was practiced is a matter of conjecture. As suggested in an earlier chapter, at first intentional selection may have been related to religious purposes. Whether or not this is so, artificial selection probably was not too long in coming after man began cultivating plants and keeping animals. Man would have realized very early that he needed only one or a few males to perpetuate his animals. We might suppose that only certain males were kept for service and that these may have been chosen for their color, the shape of their horns or the lack of horns, their docility, and so on. Later, of course, there would have been selection for milk, wool or meat production and the other types of characters that we regard as the most important today. Such selection, of course, would lead to a gradual change in the animals. The length of time required for the establishment of a new character would depend upon the nature of its inheritance and the intensity with which man managed the breeding.

Once man began to cultivate plants and keep animals, their evolution was altered. The very act of growing plants and maintaining animals away from their natural habitats could have allowed the establishment of some variants that might have perished in nature, but the new environment didn't call forth desirable mutations. Man still had to depend on chance mutation to produce change.

When man began to move his plants and animals to some distance from their original habitat, he might have brought them into areas where other wild races or closely related species were located, permitting crossing to occur naturally. Such hybridization would have allowed new combinations of genes to appear. Some of the recombinants might have been desirable from man's standpoint. Thus there was a new source of variation that would have allowed for improvement of a domesticate and the creation of new varieties.

Another source of variability and the production of new varieties would have been crossing between different strains of a domesticated species. If, as we may suppose, cattle were domesticated in different places, we might expect the different geographical strains to show some differences even though they originally came from the same wild species. Later, as man traveled with his herds or exchanged stock, opportunity would have existed for interbreeding.

Hybridization played an important role in the evolution of domesticated species, but isolation also played a very significant role. Man at times would have taken his plants and animals into areas where the related wild types were not present. Thus, for example, in some places man's newly acquired cattle would have been the only cattle, and they would thus have constituted a closed breeding population.

Such a restriction on reproduction would favor the establishment of some of the variant types, since there would be no possibility of wild-type genes entering the population kept by man. In fact, it seems likely that the most rapid evolution could have occurred by having periods of isolation followed by periods of hybridization. Some of the plants probably early provided their own method of isolation by becoming self-pollinating. Many of man's cultivated plants, including many of the cereals, are habitually self-pollinating although the wild types that gave rise to them were incapable of self-pollination and hence obligate outcrossers.

Although at this time man had no way of producing polyploid plants, it is possible that by bringing together two previously isolated species, hybrids were produced that spontaneously doubled their chromosome number to produce plants useful to man. Perhaps it was man, as previously pointed out, who brought the primitive wheats into contact with goat-grasses, permitting the formation of hybrids that gave rise to the bread wheats.

Selection continued to be man's main, if not only, method of improving his plants and animals for thousands of years. Spontaneous hybridization, as was pointed out above, doubtlessly contributed to important changes in man's domesticates. Intentional hybridization of animals must be quite ancient, but hybridization as a direct method to improve plants is fairly recent. The first well-documented plant hybrids were made in the eighteenth century, and the aim was not to improve cultivated varieties but to prove that plants reproduced sexually. Gradually intentional hybridization became used in an attempt to improve plants and with it "planning replaced accident" as C. D. Darlington has expressed it. However, until Mendel's laws were rediscovered in this century, the breeders were still working largely in the dark. Modern genetics gave plant and animal breeding a firm scientific base.

Today selection is obviously an important part of any breeding program, but in far more sophisticated ways than in past centuries. Hybridization both within and between species continues to serve as one of the principal tools of the breeder. Inbreeding followed by hybridization led to the greatly increased yields of maize. Hybridization in wheat and rice have been responsible for the development of the new "miracle" seeds that produced the green revolution. Many other plants and our principal domesticated animals have been greatly improved through hybridization.

With the artificial induction of mutations by X-rays in the late 1920's by H. J. Muller came the realization that the breeders had a potential new tool. Instead of having to depend upon nature to supply the variation, man could induce change by using X-rays and other mutagenic agents. Of course, most of the mutant types man produces are simply discarded, for most mutations—whether appearing at random in nature or as the result of a mutagenic agent—are either deleterious or not desirable for one reason or another. Efforts to use artificial mutations in plant breeding did persist, and now more than 90 commercially accepted varieties of cultivated plants are known that have resulted from induced mutations.

Nature, however, continues to be the principal source of mutations used by the breeder and will continue to be so for some time to come. The naturally occurring mutations include some very ancient ones found in older varieties of our domesticated plants and animals. Only recently has it been realized that there exists a tremendous reservoir of potentially valuable genes for use in breeding work in the old varieties of plants and animals. Although many of these varieties are rather unproductive it does not mean that they are worthless. As the new improved cereals* and other modern varieties of plants and animals spread throughout the world, the older varieties are being lost at an alarming rate. For this reason seed banks to preserve plant germ plasm have now been established. A number of years ago the Rockefeller Foundation set up a corn bank to preserve as many of the

---

*There is some danger in any one variety being widely grown for, if it should prove susceptible to a new disease, the whole crop could be virtually wiped out at one time. The southern corn blight that appeared in the United States in 1970 is an example, for the disease attacked those hybrids that had a cytoplasmic-male-sterile strain as a parent and did no damage to others. As most of the maize grown in the United States had such a parent, there was considerable loss. Thus there is an advantage to diversity, for it is unlikely that all varieties will prove equally susceptible to a disease.

Figure 10-3
Rice breeder pollinating a flower to secure hybrid seed.
(Courtesy of Rockefeller Foundation.)

Indian races of maize as possible, and various governments have taken similar steps to preserve the genetic resources of other plants. Much more needs to be done.

Various weedy species and wild species related to our domesticates still have much to contribute to breeding programs. It is up to the systematist to catalog this diversity and to indicate where it may be found. While this has been done for some groups of plants and animals, others have scarcely been touched. Botanical and zoological expeditions have been made to many parts of the earth to collect wild species and weeds but some areas have yet to be intensively explored.

A large number of man's major food plants—wheat, sugar cane, potatoes, sweet potatoes, and bananas among them—are polyploids. Just as scientists discovered how to induce mutation, they also found that polyploids could be created artificially. By treating plants with the chemical colchicine and through other methods, chromosome doubling can be induced to give rise to polyploid plants. Great hopes at one time were held for the production of economically important plants by use of this method. The hopes have not been entirely realized, but some valuable ornamentals, forage plants, and fruits, the most interesting of which perhaps is the seedless watermelon, have been developed. The understanding of polyploidy, however, has allowed tremendous improvement in some of our polyploid crops. It has, for example, allowed the transfer of desirable genes from wild diploid species of goat-grass *(Aegilops)* to the bread wheats by some most ingenious methods.

From the foregoing it should be clear that the study of evolution, genetics, systematics, in fact, of nearly all fields of biology has provided the knowledge to direct the evolution of our domesticated species. The pure research, often carried out with no direct practical applications in mind, has provided the basic knowledge for the breeder to carry out his applied research.* Few plants and animals have not been improved as the result of the breeder's efforts but much more remains to be done. We have probably not yet reached the upper limits of yields with any of our plants and animals, and some of them that serve as the basic food for many people, such as the yam, have scarcely been touched by the modern breeder. The task requires new basic research as well as more application of the knowledge now available. The new field of molecular biology holds great promise for the future, but at the same time the older methods of investigation have still not been fully exploited. For example, both morphological and systematic studies still have much to contribute to the understanding and improvement of our cultivated plants. Through a detailed study of the cacao flower and its mode of pollination, Jorge Soria of the Interamerican Institute of Agricultural Science in Costa Rica has recently come up with ways of increasing the yield of cacao,

---

*Two kinds of research may be distinguished—pure and applied. The search for a way to control a plant disease would be considered applied research because it has an immediate practical objective. The study of the evolution of a group of mosses that have no known use to man would be considered pure research. Such a study may never be of any direct practical value to mankind but, on the other hand, a hundred years from now it may prove otherwise. There is often less difficulty in getting financial support for the former study than for the latter. Yet it is the pure, or basic, research that provides the fundamental knowledge for applied research to be successful. Gregor Mendel did not set out to make a better pea or to give us the laws of genetics when he began making crosses of different types of peas. He was simply curious, and out of his curiosity came the basis of modern plant and animal breeding.

an important cash crop in many tropical countries of Africa and America.

More than a quarter of a century ago N. I. Vavilov wrote that "within the shortest period the science of breeding must advance through a whole series of upward stages to a level immeasureably higher than that on which it rests today." Great advances have been made but many more are called for. Recently William J. C. Lawrence has written that in the future "the breeder must adopt or adapt the basic strategy and tactics nature has so successfully employed in the evolution of plants, and which he ignores at his and the world's peril." Clearly the breeders and other biologists can contribute further toward eliminating hunger and malnutrition from the world, but alone they cannot solve the problem.

11

Let them eat cake?

> *It has been said, that the great question is now at issue,*
> *whether man shall henceforth start forwards with accelerated*
> *velocity toward illimitable, and hitherto unconceived*
> *improvement; or be condemned to a perpetual oscillation*
> *between happiness and misery, and after every effort remain*
> *at an still immeasurable distance from the wished for goal....*
>   *I think I may fairly make two postula.*
>   *First, that food is necessary to the existence of man.*
>   *Secondly, that the passion between the sexes is necessary,*
> *and will remain nearly in its present state....*
>   *Assuming, then, my postulata is granted, I say, that the*
> *power of population is indefinitely greater than the power in*
> *the earth to produce subsistence for man.*
>
> THOMAS ROBERT MALTHUS,
> *An Essay on the Principle of Population, 1798*

I would have liked to have concluded this book on a happy note—man's conquest of hunger and his living in harmony with his environment. Unfortunately, this is not possible, for in spite of man's many scientific advances, there are more hungry people in the present century than ever before in the world's history, primarily because there are now more people in the world than ever before. Not only are more people being born but they are staying alive longer.

The problem is largely one of the "haves" versus the "have-nots"— the "haves" being the so-called developed nations (United States, Canada, most of Europe, the Republic of South Africa, Australia, New Zealand and Japan), and the "have-nots," most of the rest of the world. A recent estimate indicates that 450 million of the world's inhabitants are well fed or even over-fed, 650 million are more-or-less adequately fed, and 2,400 million are underfed.

In the last decade or so many books have appeared on the subject— *Overcoming World Hunger, The World Food Problem, World Without Hunger, The Hungry Planet, The Black Book of Hunger, The Hungry Future, No Easy Harvest, Race Against Famine, Famine 1975,* and *Man Must Eat.* The titles of many of them are indicative of the gravity of the situation. Some of the authors are optimistic, but many are downright prophets of doom, predicting mass starvation in a few years if something isn't done now, and many think it may already be too late.

Some of the most recent books, *Seeds of Change,* for example, are more optimistic. There is perhaps some justification for this view. In 1969 the Food and Agricultural Organization of the United Nations reported that food production in 1968 increased by three percent while population increased by only two percent. This increase in food production, of course, didn't do much to erase hunger from the world, but the two statistics taken together are indeed encouraging. Then in 1969 and 1970 the new so-called "miracle seeds," the newly developed, high-yielding strains of wheat and rice, gave bumper harvests in many parts of Asia. This green revolution, as it has been popularly named, is an exciting development, but it does not signal an end to hunger and, in fact, has already created many new problems— the countries producing the grain are not equipped to store, transport, and market it. Nor even if the Asian nations become self-sufficient in grain production does it mean an immediate end to hunger in those nations, to say nothing of those countries that can not grow rice and wheat. The green revolution, in fact, holds the great danger of lulling people into a sense of complacency about the hunger problem. It is, of course, hoped that the green revolution will allow the world to buy a little time to achieve some lasting solutions of the hunger problem. Some authors, however, are predicting that the "seeds of change" may be "seeds of revolution" for those people who are not sharing in the harvests of the new seeds.

Have all of the books on hunger that have appeared in recent years served a useful purpose? We would like to think that by pointing out how critical the problem is as well as offering some possible solutions that they have, but reading about hunger makes far less impression than being hungry or seeing hungry people. And this is a large part of the problem, for most of the people who are in a position to do something about hunger have never seen a hungry person, let alone having ever experienced hunger themselves. It hardly helps to point out that dogs and cats* in the United States receive better meals than many children in Africa, or in southeastern Asia and Latin America. How much longer can the people in the wealthy nations continue to increase their standard of living while much of the world's people still do not get enough to eat? There are those who feel that nothing—or, at any rate, not enough—will be done, until catastrophe strikes and then it may be too late.

Some people in the United States have become aware that their future is endangered, that there is no universal law which says that man will be here forever. The great cause for the alarm was pollution of the environment, not hunger. The reason for their concern is obvious, of course, for pollution was all around them, and hunger, for the most part, was in some distant land or in areas of their own country from which they were isolated. It is pointless to argue which is the greater problem, pollution or hunger, for either could destroy mankind. They are, in fact, interrelated problems in many ways. It is hoped that solutions can be found for both of them. Producing more food and getting it to the people who need it sounds simpler than putting men on the moon, but the more we study the problem the more complicated we realize it is. No one with money goes hungry in any part of the world so the hunger problem is only partly a problem of not enough food; it is more a problem of getting rid of poverty. Putting men on the moon was the result of the cooperation of relatively few people of a single nation. Eliminating hunger will require the efforts of many people of many nations.

Solving the hunger problem entails getting more calories and more protein to the poor peoples of the world. Calories are not as much of a problem as are proteins, for in Chapter 7 we saw that there are many foods which will give man a full stomach and plenty of calories but which are sadly deficient in protein. The green revolution fails

---

*In 1969 more than one billion dollars was spent on pet food in the United States.

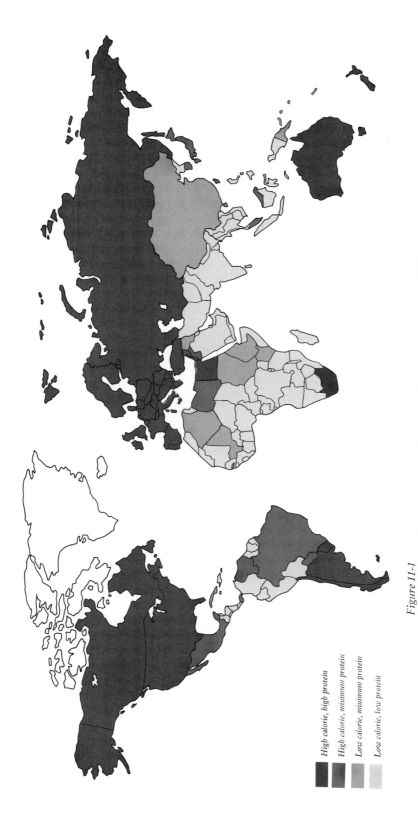

Figure 11-1
Availability of calories and protein for the world's peoples. Based on FAO data. (From Paul R. Ehrlich and Anne H. Ehrlich. *Population, Resources, Environment*, 2nd ed. W. H. Freeman and Company. Copyright © 1972.)

High calorie, high protein

High calorie, minimum protein

Low calorie, minimum protein

Low calorie, low protein

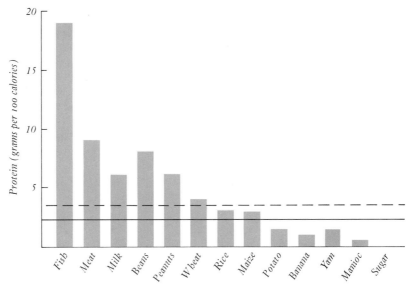

Figure 11-2
Protein-calorie ratios of various foods compared. It can be seen that some of the basic foods of tropical areas, such as banana, yam, and manioc are protein-poor. The approximate adult dietary requirement of protein is shown by the solid line and that for children by the dashed line.

somewhat in this regard, for wheat and rice alone cannot provide man with all the amino acids he needs. Meat is the best source of protein, and as we have seen it is our most expensive food. In the highly industrialized nations, man gets an average of 84 grams of protein daily, 39 of them from animal sources; in the other nations, the average intake is 52, with only 7 coming from meat. Deficiency of protein causes the disease kwashiorkor, protein deficiency and calorie deficiency together lead to marasmus. Both diseases are common in many poorer countries. The symptoms of malnutrition usually become apparent after a child leaves its mother's breast, from which it had obtained adequate protein in the form of human milk, and begins to partake of the adult diet of manioc, bananas, yams, or potatoes. If children continue to be fed a diet grossly deficient in protein, their physical growth is obstructed, they are debilitated in various ways, and, worst of all, they are likely to be mentally retarded. Thus a child

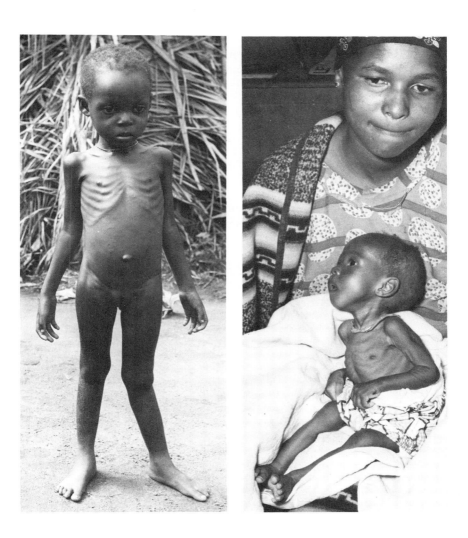

Figure 11-3
Children showing effects of malnutrition, Africa.
(Courtesy of FAO.)

kept without protein for too long can never lead a fully developed life, and this, of course, is part of the problem.

A lot has been written in recent years about our food problems. There are those who maintain that enough food is being produced to feed the world's inhabitants and that the problem is one of faulty distribution. This may to a large extent be true, but the fact is that we will need far more food in the future if we are to feed an increasing population. Various solutions have been advocated.

Bring More Land Under Cultivation

At present only about 10 percent of the land surface is under cultivation, about 17 percent is pastures or meadows, 28 percent is forested or woody areas, and about 45 percent cannot be used for agriculture because it is too steep, too rocky, too dry, too wet, or too cold. Some estimates suggest that it would be possible to increase the areas under cultivation to 30 percent but this could hardly be done in the next 50 years, if ever. Bringing more land under cultivation is not a simple matter, and the fact is that nearly all the land best suited to growing crops is already under cultivation or is occupied by cities, airports, ball parks, graveyards, and the like, and some is lost to erosion every year. This means that land that is perhaps best described as submarginal for agriculture will have to be used, which will require great effort and great cost. Some of the recent attempts in Japan and the Soviet Union at bringing new land under cultivation have shown that the return is rather small.

Some of the pasture or grassland is not suited to growing crops, being at elevations too high for cultivated plants or in areas subject to severe erosion. Much of it, of course, now serves as grazing land for livestock and is thus already being used to produce food for man. However, as we can feed about seven times as many people directly on plants as can be fed on meat, the question arises of whether we should convert the pastureland to fields for cereals and legumes. As we have seen, citizens of the wealthy nations are the meat eaters, and there would certainly be considerable resistance to giving up meat. The problem is perhaps not so much that the animals need land—in fact, with modern animal husbandry methods in the developed nations, the farm animals are utilizing less land than formerly—but that at present the rich nations are feeding their animals plant and fish

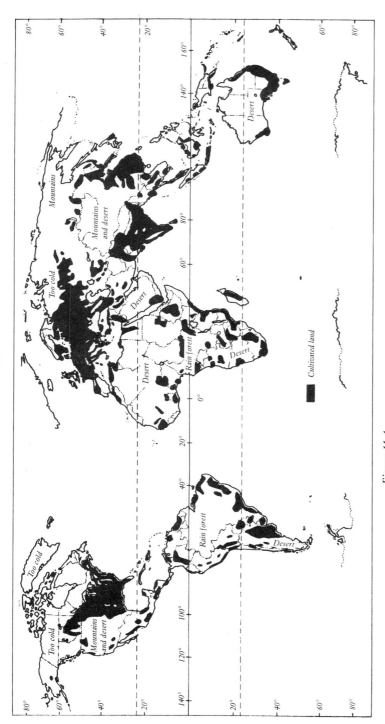

Figure 11-4
Approximate area of cultivated lands of the world.
(From Foreign Agricultural Service, USDA.)

protein that itself could go a long way toward eliminating protein deficiency among humans.* As the population grows, the question of the reduction of livestock will have to be given serious consideration. It has been suggested that the first animals to go could be sheep that are raised for wool production. Plant and synthetic fibers could readily replace wool, and the land occupied by and food given to these animals could be turned over to providing food for man. Such a practice, however, might prove more costly in the long run, for the production of synthetic fibers, like other manufactured products, requires fossil fuels, which are not a renewable resource.

Still another way to obtain more land for the production of food would be to stop growing unessential plants now being cultivated. Tobacco immediately comes to mind. People could also learn to live without tea and coffee. Much of the land used for growing these crops is not, however, the best for food plants. If, for example, tobacco were abandoned as a crop, it would create considerable unemployment and in that sense might contribute to the hunger problem. This example perhaps illustrates why the problem of hunger is so complicated. Drastic change cannot be easily made without economic disruption. The day may come, however, when the luxury crops will have to be given up in favor of essential food crops.

A few years ago people looked to the day when the vast tropical forested areas would be used for food production. It is now thought that these tropical areas are apparently best suited for growing what they have always grown—trees. When cleared, the land rapidly loses its fertility and even with the addition of fertilizer little of this land is suitable for sustained production of food crops. The advanced prehistoric cultures, with few exceptions, never developed in this kind of area and then, as now, the inhabitants of such areas depended on a slash-and-burn, shifting type of cultivation. Small areas are cleared and burned, crops are grown for a few years, and then the land is abandoned and a new area sought. Obviously, this type of cultivation will not support a large population. Although certain tree crops might be grown in these tropical forest areas, we can hardly count on them to contribute much to the world's food supply.

Nor can we count on the timber areas of the temperate zones to contribute much in the way of food. Some forested areas are hardly

---

*On the positive side it should be pointed out that urea, a nonprotein form of nitrogen manufactured from inorganic compounds, is being increasingly used in the feeding of ruminants. One-third of their nitrogen can come from this source.

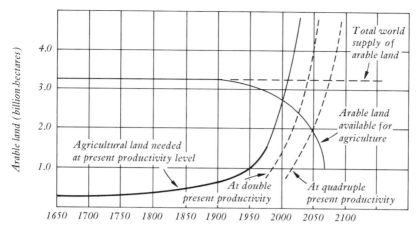

Figure 11-5
Arable land. The total world supply of arable land is about 3.2 billion hectares. About 0.4 hectares per person of arable land are needed at present productivity. The curve of land needed thus reflects the population-growth curve. The light line after 1970 shows the projected need for land, assuming that world population continues to grow at its present rate. Arable land available decreases because arable land is removed for urban-industrial use as population grows. The dotted curves show land needed if present productivity is doubled or quadrupled. (From Donella H. Meadows et al., *The Limits to Growth*. Potomac Associates and Universe Books, 1972. British and Commonwealth edition published by Earth Island Limited, London, March 1972.)

suitable for cultivation, and it might be unwise for man to destroy the timber on those areas that could be used for growing food. For one thing man desperately needs the timber resources, and moreover, these forests are not in the areas where food is most needed. In China and some other places it is now realized that clearing forests was a mistake that resulted in severely adverse ecological consequences, and some areas once cleared are now being reforested.

Some land, too wet for conventional agriculture, might be drained to make additional land available for plants but only at considerable expense, and at the same time other land, too dry for agriculture, might be brought into cultivation with irrigation. We know, of course, from success in the southwestern United States and Israel, for example, that such land can be made very productive, and, of course, many schemes are underway in various parts of the world to increase irrigation. The Aswan dam in Egypt was one such scheme, but unfortunately by the time the dam was finished the population of Egypt had so grown that the country was in exactly the same situation as before the dam was built as far as the need for food was concerned. Although we can count on some more desert land being brought into food production through irrigation, not all desert areas can be used. For one thing there isn't enough water available. Although many people may think of water as cheap and unlimited, it is not. We will, in fact, face problems with water for our cities and our industries as the population grows. Sea water can be desalinated, but it will be too expensive to use on an extensive scale and we can hardly expect it to be widely employed for agricultural purposes at least during this century.

After analyzing possible ways to bring more land into cultivation, it seems a little strange to have to report that not all good agricultural land is being utilized today. In some countries, notably the United States, some land is deliberately held out of production in order not to glut the world's markets. However, before the end of the century it seems likely that even the United States will be forced to bring all of the land possible under cultivation in order to feed people. There are also some lands in the poorer countries that are not being fully used at present. Although there has been land reform in some of these countries, there are still many large holdings that are not being used for agricultural production.

Modernize Agriculture in the Underdeveloped Nations

If new agricultural land is not going to contribute significantly to our food needs in the near future, it follows that other ways to increase food production must be pursued. One of these is to make the land now used more productive. It has been known for some time that if agriculture could be modernized throughout the world, yields could be greatly increased, but it has taken a long time to learn that the methods used in North America and Europe cannot be simply transplanted to the rest of the world.

The green revolution itself is an example of what can be done. In many parts of the world the new varieties of wheat and rice have replaced the old varieties handed down from father to son, with dramatic increases in yield. While the argument is made that people, particularly uneducated people, are slow to accept change, the green revolution proves that change can occur fairly rapidly. However, this is not to be taken to mean that there has been universal acceptances of the new varieties, for many people find their old varieties more to their taste and consequently in some regions the new rice and wheat sell for less than the traditional varieties. There are great opportunities for improving the yields of many other food plants. As has been pointed out earlier, many of the tropical crops have not been subject to modern improvement work. Unfortunately there are neither enough trained people nor enough money for much to be done along this line in many countries at the present time.

To take advantage of the high-yielding capacity of the new seeds, the farmer must use fertilizer, pesticides, herbicides, and irrigation. While there has been a great increase in the use of fertilizer in some of the developing countries, India, for example, there simply isn't enough fertilizer being made at present to take care of all the needs in these countries even if they had the money to buy it. The use of fertilizer, herbicides, insecticides, and even water can lead to severe and disastrous ecological consequences if proper care is not exercised. In fact, there is some question at present whether it is possible to use any insecticide without some unwanted side effect. It will do little for mankind to substitute more pollution for hunger. Ways must be found to increase crop yields with a minimum of damage to the environment.

Multiple cropping, or farming around the calendar by growing more than one crop a year, offers a possibility of increasing food production without finding more land. It is being practiced in some areas of the world today and more can be done in the future. Some plants are not well adapted to producing more than one crop a year. Some of the newer varieties of rice, however, having shorter growing seasons and insensitivity to day length, make it possible to produce more than one crop a year in some regions.

Part of the problem is that too many people have too little land to cultivate. In Latin America, for example, nearly half of the land belongs to 1.5 percent of the people. Even where agrarian reform has taken place, there are still many problems. In Bolivia, for example, land has been given to the campesinos, but with little administration or advice. For the most part the campesinos continue to cultivate their old plants with the same results as before. A potential danger in the land reform is that it can wreck the traditional agriculture without replacing it with anything better. There are simply not enough trained people locally available in the nonindustrialized countries to administer land reform and give advice. It should also be noted that if large efficiently run holdings are broken up into small parcels, the overall result may be a loss in food production.

The production of more food is only a part of the problem of modernizing agriculture. It has been estimated that from one-third to one-half of the crop in some regions never reaches the consumer for whom it was intended, being lost to birds, insects, rats, and various other pests, even elephants. In his book Lester Brown tells an amusing yet pathetic story. From various reports on the grain crop in India it was found that 50 percent was lost to rodents, 15 percent to cows, birds, and monkeys, 10 percent to insects, 15 percent lost in storage and transit, and 15 percent in milling and processing. Little wonder that India was hungry if 105 percent of the crop was lost! A slight exaggeration, of course, but it does emphasize not only the great need for better protection from pests, but also for better harvesting, storage, and marketing of the crops in many countries.

Modern farm machinery is, of course, being used increasingly in the tropics, but almost entirely on the large land holdings. In Latin America tractors cultivate the fields on the large estates while nearby the campesinos use oxen to plow their small holdings. The green revolution has yet to reach many small farmers. Not only do they lack modern farming equipment but they do not have the money to buy the fertilizer and other things necessary to make the new seeds productive. It should also be pointed out that many people are farming land where it would be impractical to use modern machinery even if they had the money to buy it.

With the mechanization of agriculture and more efficient production of crops throughout the world, fewer people will be needed on the farm. This will force some of the people to the cities, where, having no jobs, they will contribute to the hunger problem. Today unemployed urban dwellers are among the most underfed people in the world.

To make fertilizers and tractors and other farm implements and to provide the gasoline to run the machinery requires huge amounts of fossil fuel. Hugh Nicol in *The Limits of Man* has estimated that farmers in the wealthier countries already are spending more in the form of fossil fuel than they are getting in return from their harvests. Our fossil fuels are not unlimited, and although there is no agreement about when they will be exhausted, it is not too early to think of what consequences their depletion would have on our food production.

Food from the Oceans

Japan has been strongly dependent upon the resources of the ocean to feed her people for some time, and nutrition in the Soviet Union has been greatly improved in recent years through the greater utilization of fish. According to some current estimates, the oceans supply nearly one-fifth of the world's present consumption of protein, and most authorities agree that we can greatly increase our harvests from the sea. An optimistic view would have it tripling in the future, but doubling it seems more realistic. Whether even this can actually be done without seriously depleting the oceans' resources and how long it might take are not known. The danger already exists that our pollution of the oceans will cut into the potential yield.

Large-scale fishing industries have recently been developed in Peru and Chile. In fact, in 1970 Peru became the world's largest producer of fish. The catch fell off somewhat in 1971, and in 1972 a warm current off the coast replaced the normal cool Humboldt current that had been responsible for the large fish population. As a result the Peruvian fishing industry is seriously threatened. During the 1960's the yield from Peru alone had been estimated to be enough to erase the protein deficiency of all of South America, but the bulk of the fish was sold to the United States and Europe where it was used to feed livestock, dogs, and cats. Similarly, the shrimp industries of Mexico, Panama, and Surinam ship most of their catch to the United States. Some way must be found for those exploited nations that have a good protein source to direct it to their own people.

The increasing use of fish—both freshwater and saltwater—as human food is to be expected. At present we regularly use only 200 out of 25,000 species of fish for food. Fish aquiculture is now being expanded, although it is hardly a recent development: the Chinese were raising fish in 200 BC. The use of the so-called trash fish for fish-protein concentrates to enrich other foods can be increased. Since these fish are processed without removal of the eyes, guts, and fins, there was considerable objection in the United States a few years ago to their use as human food, which perhaps proves that no nation is without its own food "taboos."

Figure 11-6
A fish haul, mostly hake, from off the coast of Chile.
(Courtesy of FAO.)

New Food and Food Additives

In the last chapter it was pointed out that no new basic food plants and animals have been domesticated in historical times.* We have taken the plants and animals domesticated in prehistoric times, increased their yield (but not always their protein content), and continued to use them as our main foods. Our sources, as has been shown in earlier chapters, are a relatively few of the higher plants and an even smaller number of mammals and birds. Is there not some other plant waiting to be domesticated that could come to rival wheat, rice, and maize? Primitive man exploited most of the plants of his environment and largely through trial-and-error or simple luck he settled on those plants that continue to serve us. Although it cannot be stated categorically that no other plant yet to be domesticated will come to equal the foremost of those now cultivated, it seems unlikely, and few efforts are being devoted to a search for one. It would require many years of research and development to make a wild plant a productive crop, and efforts along these lines are likely to contribute little to a solution of the hunger problem. We will be better off concentrating on making our present food plants still better. As for animals there have been suggestions about potential new domesticates and some research in trying to bring a few species under domestication, but the contributions of these to future world food needs will probably be rather insignificant. As has been pointed out repeatedly, animals are relatively inefficient at converting plants to protein for man's use and man would be better off utilizing the plants directly rather than looking for new animals that would require the plants for food.

Although we may not expect any organism to come along that will rival our present cereals in supplying food, man is, however, looking for new food sources and particularly for ways in which our present food sources can be made more nutritious. Articles in the Sunday supplement sections of our newspapers and in other popular publications from time to time have hailed one or the other of the "new foods" as a solution to the world's food problem. The "new foods"

---

*A few new food plants have been domesticated in historic times, blueberries and cranberries, for example, but they do not qualify as basic. Sugar beets are also a relatively new crop, but as was pointed out previously, it was the same species of beet used by man as a food plant since prehistoric times that was developed as a source of sugar in the last century.

may play an important role in providing valuable addition to our diet and indeed, some already are, but they are not complete "solutions."

Algae and fungi have been used as food for man since the earliest times—seaweeds have long been used in the diet of Asian peoples and mushrooms and truffles have been considered delicacies by nearly all people. These multicellular algae and fungi provide little either in the way of calories or protein. Some, however, of the unicellular algae or fungi, or microorganisms, are good sources of protein. The idea of using them to manufacture protein for man goes back a good many years. The algae, particularly since they are photosynthetic organisms, can manufacture their own food from carbon dioxide using energy from the sun. It spite of all the publicity given to them, they have yet to contribute much. The procedures for growing them on a mass scale are not as simple as was once assumed. Moreover, the algae lack certain essential amino acids, and they are not easily digested by man unless the cell walls are first broken down. Their flavor is not acceptable to most people, but that, of course, can be masked in various ways.

Yeasts, which can hardly be considered a new domesticated plant, are very efficient at converting carbohydrates into protein. As early as the Second World War "meat" made partly from yeast was produced on an experimental basis. Yeasts, of course, are fungi and nonphotosynthetic, which means that they require carbohydrates and organic compounds as nutrients. Nonhuman food, such as wood pulp, can, however, be used as a source of carbohydrates for their growth. Other microorganisms can also be used to produce protein, including some bacteria that give even higher yields than do yeasts. Some are being grown using a petroleum base as a food source. Considerable effort is now being made to use microorganisms to produce protein for man, but some problems such as their purification for human use still remain. At present, single cell protein (SCP) as it is called, appears more promising as food for livestock than for humans, but, of course, such use should, theoretically at least, free more cereals and fish meal for human consumption.

Man has used seeds and various storage organs of plants to meet his principal food needs. Leaves of some plants are also eaten, as is well known, but these are hardly a major food source. Leaves contain protein, and as we have already seen in some plants, for example, manioc, they contain far more than the part traditionally eaten. Leaves also contain a lot of undigestible fiber. Man could hardly eat enough leaves to supply his protein needs, quite aside from the fact that he would find such a diet rather unattractive. But if the protein could be extracted from the leaves it could be used to supplement his diet. For several years experimental work has been carried on utilizing leaves to supply protein for man. The possibilities seem enormous, for not only could leaves from ordinary crops be used but natural stands of vegetation growing in areas unsuitable for cultivation could also be harvested for their leaves. For example, herbs from aquatic environments or trees from forests could be utilized. Although protein extraction from leaves may contribute to some extent to meeting man's needs in the future, there are tremendous problems in getting it done in the areas where it is most needed and in gaining acceptance of the resulting product as food.

In 1949 the Institute of Nutrition of Central America and Panama (INCAP) was formed and one of its most significant achievements to date has been the development of a vegetable flour, incaparina (from INCAP and harina, the Spanish word for flour), made largely from cottonseed meal. Incaparina is 25 percent or more protein. Its creation was an attempt to improve the diet of Latin Americans, who subsist mostly on maize and hence often suffer from protein deficiency. Incaparina sells at about one-fourth the cost of milk and has about the same nutritional value. In Central America and Panama about two million pounds per year were sold in the late sixties, but apparently much of this went to the middle classes. Some of the poor who most need it do not use it, either because they do not like its taste or because they are too poor to buy it.

Similar low-cost proteins from local products are now being tried in other parts of the world, but we have a long way to go before they can possibly reach all of the people who need them. A nutritious new soft drink, Vitasoy, incorporating soybean protein is now being manufactured in Hong Kong and is said to have captured one-fourth of the soft drink market there. Perhaps its introduction into the United States could improve the nutrition of our teenagers.

Most new foods are not immediately acceptable—what was true of the potato and tomato a few centuries ago still applies. People's food tastes are conservative, and this is particularly true of the uneducated, who are also, most frequently, the poorest fed. They are also the people who fail to understand proper diet. Education, of course, is needed, but at the same time we should try to increase the value of their traditional food, either through crop breeding to increase protein content, which is a slow process, or through enrichment or fortification of the standard fare without changing its taste. This is already being done in many countries—bread is enriched with thiamine, niacin, calcium, and iron. Thiamine is added to polished rice to prevent beri-beri; niacin to maize to prevent pellagra. These measures add certain essential nutrients but do not compensate for the lack of protein, which could be handled by adding the amino acids, methionine and lysine. So far this is being done only on a very limited scale.

On a recent visit to INCAP the director showed me some synthetic grains of rice being made on an experimental basis in Japan. The grains, which contain a protein supplement, were so like real rice in appearance that only an expert could ever detect that they were artificial. Added at a rate of one pound to one hundred pounds of ordinary rice they can bring the protein level to an acceptable nutritional standard, and they could be made acceptable to all rice eaters since they can be made to mimic the prevailing variety of rice in any region. But there is a long way to go before synthetic rice can be supplied to the people who need it. Someone will have to pay for it.

Although I have long ago forgotten the title, I remember vividly some scenes from a movie I saw when I was a boy. The story concerned life at some future date, and I recall that man's foods consisted of a few pills a day and that babies came from a coin machine.* The former impressed me for some reason, perhaps because I liked to eat, but I was too young to appreciate the latter. We have not reached the place where we can get all of our food in a pill, and it is unlikely that we shall ever do so. For one thing it would take a lot of pills. Moreover, man doesn't want his meals that way. Eating is an enjoyable part of life, and it is a shame that more people can't participate fully. Although our efforts to develop new foods should continue, it seems likely that those plants and animals whose domestication goes back thousands of years will continue to be our principal foods. The tremendous improvement in our plants and animals in the last quarter of a century does not mean that we have reached the limit, but hopefully should be taken to indicate that still greater advances are in store for us.

---

*I don't think there was any connection between the two.

Many Problems and Few Solutions

We shall have to bring more land into cultivation, to modernize agriculture in the underdeveloped countries, to exploit the oceans more fully, to continue to develop unconventional food sources as well as to make every effort to improve our traditional food plants and animals. All of these things require time and money. A few years ago it was estimated that half or more of the world's people are now undernourished or malnourished and one in ten is dying, directly or indirectly, from the lack of food. The world's population will double by the year 2000. If we were to do all of the things that are listed above we might be able to feed the present number of people thirty years from now but there will be twice as many people by that time. The only way to prevent greater hunger in the future is to control the birth rate. One danger in the green revolution is that it may result in a relaxation of the little effort that has been made in controlling the population growth in some of the nations where it is most needed. Practically all writers on the subject of hunger are agreed that our only hope lies in limitation of the number of people on earth, and as most people are aware, there are many other reasons for bringing our population growth as near to zero as possible. Our natural resources are simply not going to last at the rate at which they are being used today, and environmental pollution cannot be controlled unless the number of people is kept within bounds.

There are still a few who claim that something will happen to save the human race without birth control being adopted. What this will be is not spelled out, although some feel that we can manage by increasing agricultural production. The foregoing account of our agricultural potential should largely dispel such notions. In the book *Population and Food* the authors write, "attempts to adjust the number of peoples to the means of subsistence do not just insult human dignity, they constitute an evasion of the problem, rather than a solution, the effects of which could only take longer to be felt. They would mean an appalling step backward in man's long struggle to gain mastery over the created world." It is hard to understand how such a statement can be made. No one denies that we should increase production of food. We must feed the people who are already present. We are not doing it now, and the population is increasing in unprecedented numbers. We might ask, which is the greater human indignity—to see a child die of hunger or not to bring it into the world at all? Finally, it must be pointed out that much of man's present troubles result from trying to gain mastery over the earth rather than trying to live in harmony with it.

Thus far the discussion has centered on ways of producing more food, which is basically a scientific problem, but biologists and other scientists do not hold the key to solving the problem of hunger. Methods are available for producing more food from old plants as well as from new sources and for limiting the population growth. Scientists must, of course, continue to search for new and better methods, which means that research must be far better supported than at present. The problem of hunger, however, is not primarily a biological one. Poverty is the world's greatest cause of hunger, thus the problem is largely one of economics and politics. Nationalism and social attitudes, including religion, traditions, and taboos, also play important roles. Some countries of the world have been blessed with fertile land and an abundance of natural resources. They have enjoyed independence and stable governments for long periods of time and their people are educated. In contrast, the situation in the hungry countries appears almost hopeless.

Most of the hungry people, strange as it may seem, are found in the so-called agricultural nations, which simply means that they are not industrial countries. In 38 nations the average per capita annual income is $100 or less. Industrialization, some have thought, would solve the problems, but factory workers have to eat. Industrialization can't succeed until enough food is available for the workers. Modernizing agriculture requires tractors and other farm machinery and fertilizer. Unfortunately, in the past when some of these nations thought of industrialization they were more interested in automobiles than in tractors. Trained agricultural workers are needed, but few of these nations have the people or the schools for the training. A few can send students to school in North America or Europe where they may—or often may not—find educational programs that will help them in their own countries. Once they secure an education, many of the students never return to their own country. It is also sad to note that a great number of people in these countries who are in a position to get an advanced education want to become lawyers or army officers. Agricultural work has little prestige in the countries that most need it. The vast majority of the world's illiterate people are in the developing nations. Few of them understand the rudiments of good nutrition and even when means are at hand to do something about their malnutrition, they often do not. Uneducated people are less likely to try new foods or new ways of growing old ones than educated ones. Birth control, too, is more likely to be practiced by educated people. Hungry people are not energetic people—they cannot work or study effectively on a low-calorie or low-protein diet (but it hasn't seriously interfered with their reproduction). In spite of the great medical advances in this century that have kept so many people alive, many people in the tropical nations still do not have the health care that is needed. Moreover, there are many diseases to be found in the tropics that are not encountered in the temperate zones. Hungry people are more susceptible to diseases than are the well fed.

Land reform, although well developed in some countries, has scarcely started in others. Giving a person some land to farm means little unless he has good seeds, tools, and some advice on how to use them. Where new lands are available, governments often have difficulty in getting people to move. For example, many highland Indians in the Andes, although undernourished, do not want to move to the lowlands where land is available. Moreover, in many places there are no roads on which people could be moved to available new lands or their produce could be moved to markets. Many of the nations where hunger is most pronounced have one-crop economies—bananas, sugar cane, or coffee, for example. By long experience and encouragement from the industrialized nations, they have settled on one crop. In good years when the price is high they may live fairly comfortably, but when a disease strikes the plants, as is happening to coffee trees in Brazil at the moment, or when there is a drop in world prices, there is great suffering. Why don't these people grow food for their own use instead of crops for export? Simply because only through exports can they get much needed cash. Theoretically at least this should allow them to buy more food than they could have grown on their own and, they may hope, industrial goods from other nations. Somehow it rarely works this way. Crop diversification, of course, has long been called for, but like so many remedies, it is more easily said than done. A transitional period is necessary to effect the conversion and this requires the investment of money and technical knowledge.

Too often the hungry nations have suffered from corrupt, inefficient and unstable governments. In some countries the wealthy people or those in power do not want change and have successfully fought reforms designed to improve the lot of their people, often with encouragement from the more powerful nations. Even when an honest, efficient government takes over, it may not last long because the people may be so impatient for change that the new government is given little opportunity to make any real accomplishments. Thus long-range programs that might have some chance of success may never get off the ground.

The list of the ills of the poorer nations could be increased but this account should serve to illustrate the magnitude of the problem. Obviously the situation varies from one country to another. Some countries have made progress with some of their problems but any progress that has been made can be virtually wiped out in a very short time by an increase in population. There are, of course, several ways in which these countries could go about putting their own houses in order, but for the most part they are trapped in a self-perpetuating crisis of poverty, malnutrition, illiteracy, and population growth. These nations would like to solve their own problems, but obviously they cannot. They need massive help from the outside.

This has, of course, been known for well over a quarter of a century and help has been given—but the help has not always been enough nor of the right kind. Some have argued that only long-term projects undertaken on an international scale could be really effectual. There is, of course, an international body, the Food and Agricultural Organization of the United Nations, dedicated to the task of solving the world's food problem. It was at one time proposed that nations give 1 percent of their earnings to combat world hunger but this has never materialized. FAO has made significant contributions but it does not have the financial backing to do the job. As long as the world's major powers are engaged in a cold war there seems little prospect that international cooperation will succeed. The amount now spent on armaments and space research might not solve the problem, but it could go a long way toward doing so. As this is being written the United States is still engaged in a long drawn-out war in Viet Nam, Catholics are fighting Protestants in Ireland, the Near East is in turmoil and active conflict between Israel and Egypt looms as a distinct possibility, and a bloody civil war in Pakistan followed by war with India has created grave problems in that part of the world, an area where hunger was already pronounced. Only a few writers on hunger have made the point that the elimination of war could be one of the greatest contributions toward solving world hunger, perhaps because of a feeling that this is too much to hope for.

Not only is there need for the improvement of the political climate of the world but of the economic one also. As Georg Borgstrom has put it, "Maybe the world's economy has become too important an issue to be handled by economists." He was being only half facetious. The world's economy is obviously controlled by the highly industrialized nations, and it appears that the other nations can never hope to become a member of the club until they can feed themselves—and they can never feed themselves until they become developed nations. Little wonder that the problem appears almost hopeless.

What should be the future role of the United States? During the 1960's, food aid from the United States was feeding 100 million people a year and only through this aid was severe famine prevented in India. As some unsympathetic foreigners are quick to point out, what else was the United States to do with a huge surplus of wheat? And cynics have stated that by keeping so many people alive the United States only contributed to future world hunger. The aims, of course, were not wholly unselfish, for there was not only a hope that it would get struggling nations on their feet, but also that it would make them friendly allies and future economic consumers. The aims weren't always realized. The days of the huge surpluses are over in the United States, and farmers are paid not to plant for fear of serious economic consequences if there were a glut on the world's markets. Some authorities are recommending the same approach for other countries in order to advance the agricultural revolution. This may seem strange when people are hungry in parts of the world, but it is a fact of economics. It costs money to give away food, and the United States is the only country that has engaged in it to any great extent in the past. Other nations in a position to contribute—Canada and Australia, for example—have supplied little. Many people in the United States like to think that their country is contributing to solving the problems of the world and can't understand why other nations regard us with suspicion or don't show appreciation of our aid. The adverse opinions of aid recipients can be partly explained by the fact that it is often natural to feel resentment upon receiving charity, but more than that is involved. It is somewhat saddening but perhaps enlightening to read some of the comments made by foreign writers on the subject of the United States' aid, comments to the effect that the United States is worried only about protecting itself, that aid is used as a bribe, that it goes only to those who least need it and often strengthens vested interests, and, perhaps the most cruel criticism of all, that the motive underlying every "gift" is related to the United States' efforts to accumulate wealth. Is the last statement to be credited to "sour grapes" on the part of nations envious of the wealth of the United States or is it the truth? There are, of course, nearly always strings attached to foreign aid, not only from the United States but from other countries as well.

Although, as Gunnar Myrdal has expressed it, there is a feeling in the poorer countries that insincerity and hypocrisy are pervasive in all Western nations, the sentiments expressed above help to illustrate why an international undertaking might have greater hope of success than that of any single nation. As an international undertaking seems unlikely at present, however, the United States must continue to do as much as it can alone.

Many Americans are opposed to giving any form of foreign aid. Some are in favor of giving only military aid—and that only in instances in which they think our safety is involved. Obviously there are many critical demands being made for money at home—for programs to reduce the pollution of our environment, to correct the problems of the cities, to feed our own hungry people, in addition to those that provide education, build roads, and carry on the many other functions that a government must perform. Given the domestic demands, great effort would be required to persuade people that more of the tax dollar should go to support other nations. The majority of American voters will probably never favor foreign aid from purely altruistic motives. Therefore, foreign-assistance programs must continue to be sold to the American people on the basis of self-interest, to speak bluntly. Clifford M. Hardin, the first secretary of agriculture in the Nixon administration, gives four reasons for foreign aid. (1) It will reduce the chance of widespread famine and suffering, which could make huge demands on the United States. (2) It can insure rapid economic growth in developing nations, which would be to our benefit. (3) It will promote a self-sustaining economic growth that can lower world tensions and promote peace. (4) It can develop societies more congenial to ours.

Although even some of these motives may be unacceptable to certain Americans for particular reasons, they probably form the only basis for which sufficient support can be engendered to increase foreign assistance. It is hoped that the greatest part of foreign-assistance money would not go for military aid as it has in the past, but for economic, educational, and agricultural development. The latter should not be in the form of gifts of food, but in technical assistance, education, and loans. It is now realized that direct food aid probably hinders agricultural development in the country that receives it. Some of the technical help from the United States in the past has also not been too helpful, sometimes simply because the most capable people have not been sent, but there have been some successes. In fact, in this century the United States has contributed more to the world's agricultural development through research and foreign aid than has any other country. However, if any one organization were to be singled out for making making contributions to solving world hunger it would probably not be any agency of the government but the Rockefeller Foundation. There are probably three reasons for this foundation's great success—it comprises a group of able and dedicated scientists, the efficiency of its operation is seldom if ever equalled by any governmental agency, and it is more often accepted in other countries, for its motives, unlike those of a governmental agency, are not always suspect. The Rockefeller Foundation, although operating on a fairly generous budget, does not have the financial resources to do more than a limited amount.

The Paddocks called their book, *Famine 1975! America's Decision: Who Will Survive*. Although the famine they envision will probably not strike by 1975, largely because of the green revolution, the rest of their title may well still apply for the future.

What will be the position of man in the year 2000? Will pollution, poverty, and hunger be rampant? Will we have conquered major problems, permitting all the peoples of the world to enjoy the high living standards now found in a few nations? Or will the situation be about the same as it is now—but with twice as many people on earth, about half of whom will be undernourished or malnourished and some of whom will be dying of hunger?

References

GENERAL

Anderson, Edgar, 1952. *Plants, Man and Life*. Little, Brown, Boston. (Written for the interested layman, a rather unconventional but fascinating introduction to weeds, cultivated plants, and the botanists who study them. Some of the material regarding the origin of certain crops has been superseded by recent discoveries. Available as a paperback from the University of California Press, Berkeley)

Baker, H. G., 1970. *Plants and Civilization*, 2nd ed. Wadsworth, Belmont, Calif. (An abbreviated account of the most important economic plants)

Economic Botany, the Journal of the Society for Economic Botany, published by the New York Botanical Garden, Bronx, N.Y. (A journal of applied botany and plant utilization)

Hutchinson, Sir Joseph (ed.), 1965. *Essays on Crop Plant Evolution*. Cambridge University Press, Cambridge, England. (A collection of papers by botanical specialists dealing with the beginnings of agriculture in Northwest Europe, maize, sorghum, temperate-zone cereals, potatoes, forage grasses, and concluding with a general discussion by the editor, one of the world's foremost authorities on the subject)

Janick, Jules, Robert W. Schery, Frank W. Woods, and Vernon W. Ruttan, 1969. *Plant Science: An Introduction to World Crops*. W. H. Freeman and Company, San Francisco. (An introductory text covering botanical, technological, and economic aspects of agriculture)

Janick, Jules, Robert W. Schery, Frank W. Woods, and Vernon W. Ruttan (eds.), 1970. *Plant Agriculture; Readings from Scientific American*. W. H. Freeman and Company, San Francisco. (Reprints of semipopular articles by specialists dealing with agricultural origins, plant growth and development, the environment of the plant, production technology, and food needs and potentials)

Purseglove, J. W., 1968–   . *Tropical Crops: Dicotyledons,* 2 vols. *Tropical Crops: Monocotyledons,* 2 vols. Wiley, New York. (A treatment of virtually all tropical economic plants, many of which, of course, are also cultivated in the temperate zones. In addition to detailed descriptions and many illustrations, there are remarks on pollination, germination, propagation, chemical composition, husbandry, pests and diseases, breeding, and origins)

Riley, Carroll L., J. Charles Kelley, Campbell W. Pennington, and Robert L. Rands (eds.), 1971 *Man across the Sea: Problems of Pre-Columbian Contacts.* University of Texas Press, Austin. (A series of papers dealing with many aspects of the "diffusion versus independent development controversy." Papers are included on chickens, coconut, bottle gourd, sweet potato, maize, squash, and beans)

Sauer, Carl O., 1969. *Seeds, Spades, Hearths, and Herds: The Domestication of Animals and Foodstuffs,* 2nd ed. M.I.T. Press, Cambridge, Mass. (Much of this book is a reprint of the original edition published in 1952 under the title *Agricultural Origins and Dispersals* and hence does not take into account new information now available. Nevertheless it remains one of the most stimulating discussions of the origin of agriculture)

Schery, Robert W., 1972. *Plants for Man,* 2nd ed. Prentice-Hall, Englewood Cliffs, N. J. (An account of economic plants, including those used for food as well as in other ways)

Stakman, E. C., Richard Bradfield, and Paul C. Mangelsdorf, 1967. *Campaigns Against Hunger.* Harvard University Press, Cambridge, Mass. (An account of the work of the Rockefeller Foundation. Includes discussion on improvement of cereals, beans, potatoes, and livestock)

Struever, Stuart (ed.), 1970. *Prehistoric Agriculture.* Natural History Press–Doubleday, Garden City, N. Y. (A collection of papers from various sources treating the origin of agriculture, the consequences of early agriculture in various parts of the world, the role of agriculture in the development of civilization, and the early history of the domestication of various plant and animal groups)

Ucko, Peter J., and G. W. Dimbleby (eds.), 1969. *The Domestication and Exploitation of Plants and Animals.* Aldine, Chicago. (Original papers by fifty scientists dealing with virtually all aspects of domestication. Topics covered include the origin of domestication, methods of investigation, and human nutrition, as well as the treatment of many different plants and animals. Although the Old World receives far more attention than the New World and animals more attention than plants, the book is nevertheless the best source book on the subject of domestication)

CHAPTER 1

Braidwood, Robert J., 1967. *Prehistoric Men*, 7th ed. Scott, Foresman, Glenview, Ill.

Byers, Douglas S. (ed.), 1967. *The Prehistory of the Tehuacan Valley*, vol. 1, *Environment and Subsistence*. University of Texas Press, Austin.

Flannery, Kent V., 1965. The ecology of early food production in Mesopotamia. *Science 147*: 1247–1256.

Gorman, Chester F., 1969. Hoabinhian: a pebble-tool complex with early plant associations in Southeast Asia. *Science 163*: 671–673.

Heiser, C. B., 1965. Cultivated plants and cultural diffusion in nuclear America. *American Anthropologist 67*: 930–949.

Helbaek, Hans, 1959. Domestication of food plants in the Old World. *Science 130*: 365–372.

Issac, Erich, 1970. *Geography of Domestication*. Prentice-Hall, Englewood Cliffs, N. J.

Lee, Richard B., and Irven de Vore (eds.) with assistance of Jill Nash, 1968. *Man the Hunter*. Based on a symposium. Aldine, Chicago.

MacNeish, Richard S., 1964. Ancient mesoamerican civilization. *Science 143*: 531–537.

Pickersgill, Barbara, 1969. The archaeological record of chili peppers (*Capsicum* spp.) and the sequence of plant domestication in Peru. *American Antiquity 34*: 55–61.

Towle, Margaret A., 1961. *The Ethnobotany of Pre-Columbian Peru*. Viking Fund Publication in Anthropology No. 30. Wenner-Gren Foundation, New York.

Wright, H. E. J., 1970. Environmental change and the origin of agriculture in the Near East. *BioScience 20*: 210–212.

CHAPTER 2

Allen, Grant, 1894. The origin of cultivation. *The Fortnightly Review 61*: 578–592.

Coe, Michael D., and Kent V. Flannery, 1964. Microenvironments and mesoamerican prehistory. *Science 143*: 650–654. (Reprinted in Struever, see General References)

Flannery, Kent V., 1969. Origins and ecological effects of early domestication in Iran and the Near East. Pp. 73–100 *in* Ucko and Dimbleby, see General References. (Reprinted in Struever, see General References)

Gaster, T. H. (ed.), 1964. *Sir James Frazer's The New Golden Bough*. Mentor, New York. (An abridged and somewhat revised edition of a classic work. Probably the best source on myth, magic, and religion in relation to primitive and ancient agriculture)

Hatt, Gudmund, 1951. The corn mother in America and in Indonesia. *Anthropos 46:* 853–914.

Heiser, C. B., 1969. Some considerations of early plant domestication. *BioScience 19:* 228–231.

Issac, Erich, 1962. On the domestication of cattle. *Science 137:* 195–204. (Reprinted in Struever, see General References)

James, E. O., 1959. *The Cult of the Mother-Goddess.* Praeger, New York.

James, E. O., 1960. *The Ancient Gods.* Putnam's, New York.

James, E. O., 1962. *Sacrifice and Sacrament.* Barnes and Noble, New York.

James, E. O., 1966. *The Tree of Life.* Brill, Leiden.

Simoons, Frederick J., 1968. *A Ceremonial Ox of India.* University of Wisconsin Press, Madison.

CHAPTER 3

Barnicot, N. A., 1969. Human nutrition: evolutionary perspectives. Pp. 525–529 *in* Ucko and Dimbleby, see General References.

Brothwell, D. R., 1969. Dietary variation and the biology of earlier human populations. Pp. 531–545 *in* Ucko and Dimbleby, see General References.

Lappé, Frances M., 1971. *Diet for a Small Planet.* Ballantine, New York. (A book about plant protein for human use)

Pyke, Magnus, 1970. *Man and Food.* World University Library, Mc-Graw-Hill, New York. (An excellent and interesting treatment of human nutrition)

Simoons, Frederick J., 1961. *Eat Not This Flesh.* University of Wisconsin Press, Madison. (Origin of food taboos)

Yudkin, John, 1969. Archaeology and the nutritionist. Pp. 547–552 *in* Ucko and Dimbleby, see General References.

CHAPTER 4

Cole, H. H. (ed.), 1966. *Introduction to Livestock Production, Including Dairy and Poultry,* 2nd. ed. W. H. Freeman and Company, San Francisco.

Cranstone, B. A. L., 1969. *Animal Husbandry: the Evidence from Ethnography.* Pp. 247–264 *in* Ucko and Dimbleby, see General References.

Gilmore, R. M., 1950. Fauna and ethnozoology of South America. Pp. 345–464 *in* Julian H. Steward (ed.), *Handbook of South American Indians.* Smithsonian Institution, Bureau of American Ethnology, Washington, D.C. (Smithsonian Institution Bulletin 143)

Hole, Frank, Kent V. Flannery, and James A. Neely, 1969. *Prehistory and Human Ecology of the Deh Luran Plain: An Early Village Sequence from Khuzistan, Iran.* University of Michigan, Museum of Anthropology, Ann Arbor. (Excerpts reprinted in Struever, see General References)

Leeds, Anthony, and Vayda, Andrew P., 1965. *Man, Culture and Animals: the Role of Animals in Human Ecological Adjustments.* American Association for the Advancement of Science, Washington, D.C.

Perkins, Dexter, Jr., 1969. Fauna of Çatal Hüyük: evidence for early cattle domestication in Anatolia. *Science 164:* 177–179.

Reed, Charles A., 1969. Animal domestication in the prehistoric Near East. *Science 130:* 1629–1639. (Reprinted in Struever, see General References.

Reed, Charles A., 1969. The pattern of animal domestication in the prehistoric Near East. Pp. 361–380, *in* Ucko and Dimbleby, see General References.

Zuener, Frederick E., 1963. *A History of Domesticated Animals.* Hutchinson, London. (A very thorough treatment of all of the domesticated animals)

CHAPTER 5

WHEAT

Harlan, Jack, 1967. A wild wheat harvest in Turkey. *Archaeology 20:* 197–201.

Helbaek, Hans, 1966. 1966—Commentary on the phylogenesis of *Triticum* and *Hordeum*. *Economic Botany 20:* 350–360.

Peterson, R. F., 1965. *Wheat.* Interscience, New York.

Reitz, Louis P., 1970. New wheats and social progress. *Science 169:* 952–955.

Riley, R., 1965. Cytogenetics and the evolution of wheat. Pp. 103–122 *in* Hutchinson, see General References.

Zohary, Daniel, 1969. The progenitors of wheat and barley in relation to domestication and agricultural dispersal in the Old World. Pp. 47–66 *in* Ucko and Dimbleby, see General References.

Zohary, Daniel, 1971. Origin of South-West Asiatic cereals: wheat barley, oats and rye. Pp. 235–260 *in* Peter H. Davis, Peter C. Harper, and I. C. Hedge (eds.)., *Plant Life of South-West Asia.* Botanical Society of Edinburgh, Edinburgh.

RICE

Chandraratna, M. F., 1964. *Genetics and Breeding of Rice.* Longmans, London.

Ghose, R. L. M., M. B. Ghatge, and V. Subrahmanyan, 1960. *Rice in India*. Indian Council of Agricultural Research, New Delhi.

Grist, D. H., 1965. *Rice*, 4th ed. Longmans, London.

Oka, Hiko-Ichi, and H. Morishima, 1967. Variations in the breeding system of a wild rice, *Oryza perennis*. *Evolution 21:* 249–258.

Steeves, Taylor, 1952. Wild rice—Indian food and a modern delicacy. *Economic Botany 6:* 107–142.

MAIZE

de Wet, J. M. J., and J. R. Harland, 1972. Origin of maize: the tripartite hypothesis. *Euphytica 21:* 271–279.

Galinat, Walton C., 1971. The origin of maize. *Annual Review of Genetics 5:* 447–478.

Mangelsdorf, P. C., Richard S. MacNeish, and Walton C. Galinat, 1964. Domestication of corn. *Science 143:* 538–545.

Wallace, Henry A., and William L. Brown, 1956. *Corn and Its Early Fathers*. Michigan State University Press, East Lansing. (A historical treatment of the development of hybrid corn. Henry A. Wallace was a vice-president under Franklin D. Roosevelt)

Weatherwax, Paul, 1954. *Indian Corn in Old America*. Macmillan, New York. (A good introduction to the corn plant and its relatives, containing detailed consideration of uses of corn among the Indians. Excellently illustrated)

SUGAR CANE

Barnes, A. C., 1964. *The Sugar Cane*. Interscience, New York.

Deerr, Nöel, 1949. *The History of Sugar*, 2 vols. Chapman and Hall, London.

CHAPTER 6

Gentry, Howard Scott, 1969. Origin of the common bean, *Phaseolus vulgaris*. *Economic Botany 23:* 55–69.

Hymowitz, T., 1970. On the domestication of the soybean. *Economic Botany 24:* 408–421.

Kaplan, Lawrence, 1965. Archaeology and domestication in American *Phaseolus* (beans). *Economic Botany 19:* 358–368. (Reprinted in Struever, see General References)

Krapovickas, A., 1969. The origin, variability and spread of the ground nut *(Arachis hypogaea)*. Pp. 427–441 *in* Ucko and Dimbleby, see General References.

Whyte, R. O., G. Nilsson-Leissmer, and H. C. Trumble, 1953. *Legumes in Agriculture*. Food and Agricultural Organization of the United Nations, Rome. (FAO Agricultural Studies No. 21)

CHAPTER 7

Coursey, D. G., 1967. *Yams.* Longmans, London.

Greenwell, Amy B. H., 1947. Taro—with special references to its cultures and uses in Hawaii. *Economic Botany 1:* 276–289.

Hawkes, J. G., 1967. The history of the potato. *Journal of the Royal Horticultural Society 92:* 207–224, 249–262, 288–302.

Jones, William O., 1959. *Manioc in Africa.* Stanford University Press, Stanford, Calif.

Nishiyama, Ichizo, 1971. Evolution and domestication of the sweet potato. *Botanical Magazine* (Tokyo) *84:* 377–387.

O'Brien, Patricia J., 1972. The sweet potato: its origin and distribution. *American Anthropologist 74:* 342–365.

Rogers, David J., 1965. Some botanical and ethnological considerations of *Manihot esculenta. Economic Botany 19: 369–377.*

Salaman, R. N., 1949. *The History and Social Influence of the Potato.* Cambridge University Press, London.

Simmonds, N. W., 1966. *The Evolution of the Bananas,* 2nd ed. Longmans, London.

Simmonds, N. W., 1969. *Bananas.* Longmans, London.

Taylor, Norman, 1965. *Plant Drugs that Changed the World.* Dodd, Mead, New York. (see for medicinal uses of *Dioscorea*)

Ugent, Donald, 1970. The potato. *Science 170:* 1161–1166. (Primarily concerned with origin of the potato)

CHAPTER 8

Child, Reginald, 1964. *Coconuts.* Longmans, London.

Nixon, Roy W., 1951. The date palm, "tree of life" in the subtropical deserts. *Economic Botany 5:* 274–301.

Purseglove, J. W., 1968. The origin and distribution of the coconut. *Tropical Science 10:* 191–199.

Sauer, Jonathan D., 1971. A reevaluation of the coconut as an indicator of human dispersal. Pp. 309–319 *in* Riley *et al.,* see General References.

CHAPTER 9

Amerine, M. A., and V. L. Singleton, 1968. *Wine.* University of California Press, Berkeley.

Bailey, Liberty Hyde. 1949. *Manual of Cultivated Plants,* 2nd. ed. Macmillan, New York. (Descriptions and keys for identification of all plants commonly cultivated in the United States and Canada)

Heiser, Charles B., 1969. *Nightshades, the Paradoxical Plants.* W. H. Freeman and Company, San Francisco. (Includes discussion of tomato, eggplant, and other solanaceous plants)

Herklots, G. A. C., 1972. *Vegetables in Southeast Asia.* Allen and Unwin, London.

Nieuwhof, M., 1969. *Cole Crops.* Hill, London.

Reuther, W., L. D. Batchelor, and H. J. Webber (eds.)., 1967. *The Citrus Industry,* vol. 1. *History, World Distribution, Botany, and Varieties,* Rev. ed. Division of Agricultural Science, University of California, Berkeley.

Rosengarten, Frederic, 1969. *The Book of Spices.* Livingston, Wynnewood, Pa.

Whitaker, Thomas W., and Glen N. Davis, 1962. *Cucurbits: Botany, Cultivation and Utilization.* Interscience, New York.

Woodroof, Jasper G., 1967. *Tree Nuts: Production, Processing, Products,* 2 vols. Avi, Westport, Conn.

CHAPTER 10

Darlington, C. D., 1963. *Chromosome Botany, and the Origins of Cultivated Plants,* 2nd ed. Hafner, New York.

De Candolle, A. P., 1886. *Origin of Cultivated Plants.* Appleton, New York.

Frankel, O. H., and E. Bennett (eds.), 1970. *Genetic Resources in Plants—their Exploration and Conservation.* Blackwell, Oxford.

Johansson, I., and J. Rendel, 1968. *Genetics and Animal Breeding.* W. H. Freeman and Company, San Francisco.

Lawrence, William J. C., 1968. *Plant Breeding.* Arnold, London.

Lerner, Isadore M., and H. P. Donald, 1966. *Modern Developments in Animal Breeding.* Academic Press, London.

Schwanitz, Franz, 1966. *The Origin of Cultivated Plants.* Harvard University Press, Cambridge, Mass.

Smith, H. H., 1971. Broadening the base of genetic variability in plants. *Journal of Heredity* 62: 265–276.

Stebbins, G. L., Jr., 1971. *Processes of Organic Evolution,* 2nd. ed. Prentice-Hall, Englewood Cliffs, N. J. (An excellent general treatment of the subject for the beginning student)

CHAPTER 11

Aldrich, Daniel G., Jr., 1970. *Research for the World Food Crisis.* American Association for the Advancement of Science, Washington, D. C.

Bennett, Ivan L., Jr., 1969. Problems of world food supply. In *All Congress Symposium, World Food Supply.* International Botanical Congress, Seattle, Washington.

Borgstrom, Georg, 1969. *Too Many.* Macmillan, New York.

Borgstrom, Georg, 1972. *The Hungry Planet,* 2nd ed. Macmillan, New York.

Brown, Lester R., 1970. *Seeds of Change*. Praeger, New York.

Brown, Lester R., and G. W. Finsterbusch, 1972. *Man and His Environment: Food*. Harper and Row, New York.

Bunting, A. H. (ed.), 1970. *Change in Agriculture*. Duckworth, London.

Castro, Josué de, 1967. *The Black Book of Hunger*. Translated by Charles Lam Markmann. Funk and Wagnalls, New York.

Cépède, Michel, François Houtart, and Linus Grond, 1964. *Population and Food*. Sheed and Ward, New York.

Dumont, René, and Bernard Rosier, 1969. *The Hungry Future*. Praeger, New York. (A translation from the French)

Ehrlich, Paul R., and Anne H. Ehrlich, 1972. *Population, Resources, Environment: Issues in Human Ecology*, 2nd. ed. W. H. Freeman and Company, San Francisco.

Food and Agricultural Organization of the United Nations, 1970. *The State of Food and Agriculture 1970*. FAO, Rome.

Freeman, Orville L., 1968. *World Without Hunger*. Praeger, New York.

Hardin, Clifford M., 1969. *Overcoming World Hunger*. Prentice-Hall, Englewood Cliffs, N. J.

Hopcraft, Arthur, 1968. *Born to Hunger*. Houghton Mifflin, Boston.

Myrdal, Gunnar, 1970. *The Challenge of World Poverty: a World Anti-poverty Program in Outline*. Pantheon, New York.

Paddock, William C., and Paul Paddock, 1967. *Famine 1975! America's Decision: Who Will Survive*. Little, Brown, Boston.

Paddock. William C., 1970. How green is the green revolution? *BioScience 20:* 897–902.

Pirie, N. W., 1969. *Food Resources: Conventional and Novel*. Penguin Books, London.

Stamp, Elizabeth, 1967. *The Hungry World*. E. J. Arnold and Sons, Leeds.

Thurston, H. David, 1969. Tropical agriculture: a key to the world food crisis. *BioScience 19:* 29–34. (Good treatment of the problems of education in agriculture for the tropical nations)

Index